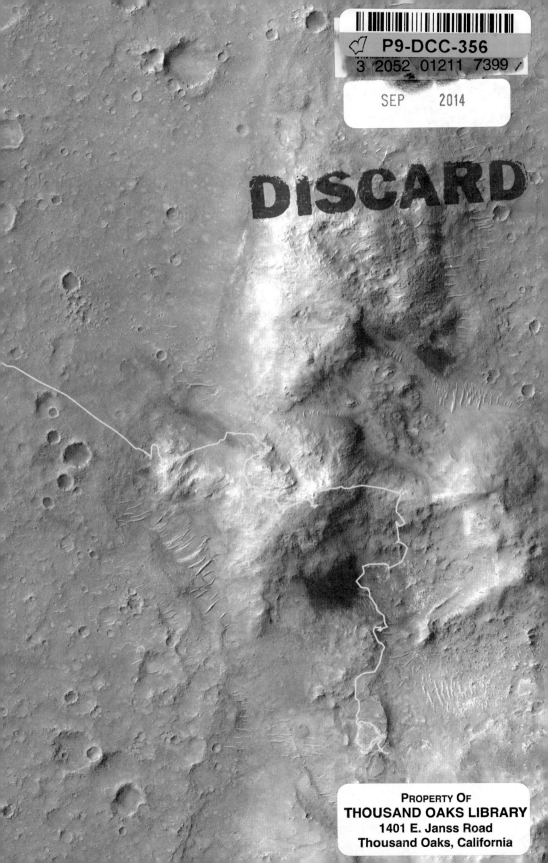

COLLECTION MANAGEMENT

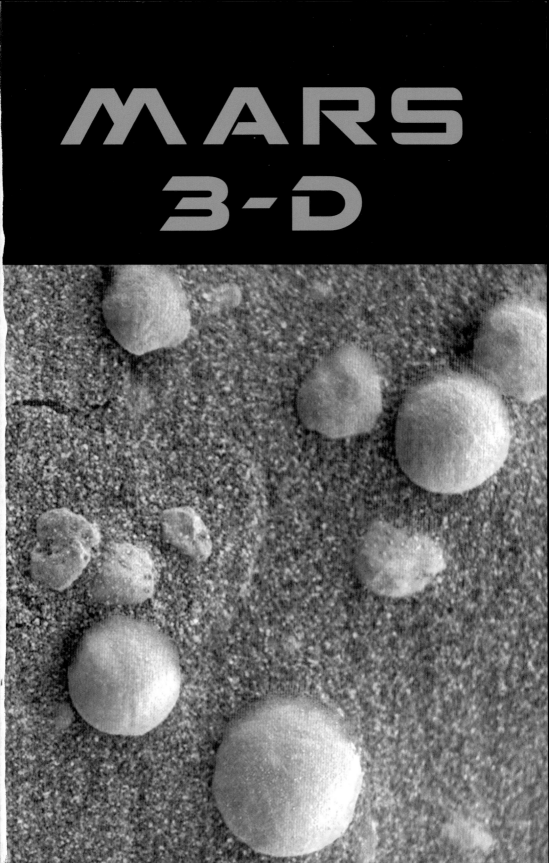

MARS
3-D

MARS
3-D

A Rover's-Eye View
of the Red Planet

JIM BELL

STERLING
New York

STERLING
New York

An Imprint of Sterling Publishing
387 Park Avenue South
New York, NY 10016

This revised and expanded edition of Mars 3-D published in 2014

ISBN 978-1-4549-1178-4

Library of Congress Cataloging-in-Publication Data

Bell, Jim, 1965-
    Mars 3-D : a rover's-eye view of the red planet / Jim Bell. -- Revised and expanded edition.
        pages cm
    Includes index.
    ISBN 978-1-4549-1178-4
    1. Mars (Planet)--Exploration. 2. Mars (Planet)--Pictorial works. 3. Three-dimensional imaging
in astronomy. I. Title. II. Title: Mars three-D.
    QB641.B445 2014
    523.43--dc23
                                    2013030701

Distributed in Canada by Sterling Publishing
℅ Canadian Manda Group, 165 Dufferin Street
Toronto, Ontario, Canada M6K 3H6
Distributed in the United Kingdom by GMC Distribution Services
Castle Place, 166 High Street, Lewes, East Sussex, England BN7 1XU
Distributed in Australia by Capricorn Link (Australia) Pty. Ltd.
P.O. Box 704, Windsor, NSW 2756, Australia

For information about custom editions, special sales, and premium and corporate purchases,
please contact Sterling Special Sales at 800-805-5489 or specialsales@sterlingpublishing.com.

Manufactured in China

2 4 6 8 10 9 7 5 3 1

www.sterlingpublishing.com

Opposite the title page: *Viking Orbiter* mosaic view of thin clouds and hazes above Argyre Basin, in
the heavily cratered southern highlands.   523.43

All Photos courtesy of NASA/JPL except:
NASA/J. Bell: pages ii-iii
NASA/G. Legg: pages 18
NASA/JPL/Cornell: pages ix, 9, 17, 27, 38, 39, 40, 42, 44, 45, 46, 48, 49, 50, 51, 52, 53, 54, 55, 56, 57,
58, 60, 61, 62, 63, 64, 65, 66, 67, 68, 69, 70, 72, 73, 74, 75, 76, 77, 88, 89, 90, 92, 94, 95, 96, 98, 99,
100, 102, 104, 105, 106, 107, 108, 110, 111, 112, 113, 114, 115, 116, 117, 118, 119, 120, 121, 122, 123,
124, 126, 127, 128, 129, 130, 131, 142, 143, 144, 145
NASA/JPL/Cornell/ASU: pages 78, 79, 80, 81, 82, 83, 84, 85, 132, 133, 135, 136, 137, 138, 139
NASA/JPL/Univ. Arizona: pages 86, 87, 134, 140, 141
NASA/JPL/USGS: pages i, 31, 41, 47, 53, 59, 71, 91, 93, 97, 101, 103, 109, 125, 135
NASA/JPL/MSSS: pages 142, 144, 145, 146, 148, 149, 150, 151, 152, 153, 154, 156, 158, 159, 160, 161
NASA/JPL/MSSS/Fred Konkin: page 143
NASA/JPL/MSSS/Ed Truthan: page 147

CONTENTS

ACKNOWLEDGMENTS

THE MARS ROVERS *Spirit, Opportunity,* and *Curiosity* were conceived long before they were born, by small groups of scientists, engineers, and managers who share a remarkable vision: that of being able to project themselves onto Mars through robotics, and then to explore the place remotely as if they were there in person. These colleagues, and the subsequent team of thousands of incredibly talented people from around the world who came together to carry out this vision, are the ones who deserve the biggest acknowledgment here: for bringing these rovers "to life" and for enabling me to share with everyone the remarkable photographic views showcased in this book. I am especially grateful to the engineers and technicians at NASA's Jet Propulsion Laboratory (JPL) in Pasadena, California, and Malin Space Science Systems, Inc. (MSSS) in San Diego, California who designed and painstakingly assembled the Mars rover cameras to specific and exacting tolerances (enabling the high-quality, stereoscopic views depicted here). I would also like to thank the science and engineering team members who have been helping to write the sequences and to acquire the photos from *Spirit, Opportunity,* and *Curiosity.* Because of the long lifetimes of these rovers, and their daring but highly successful pinpoint landing systems, our hopes were fulfilled, and indeed surpassed, in a most dramatic fashion!

Fusing left and right camera images into 3-D anaglyphs is a tricky business, and it depends greatly on the proper processing and geometric projection of the images. In that regard, I am very grateful to Jonathan Joseph at Cornell University for developing the extremely capable and flexible Mermap and Merview image processing software, which allowed me to generate many of the 3-D as well as color views shown here, and to the teams at MSSS and the JPL Multimission Image Processing Laboratory (especially Bob Deen and Kris Capraro) for generating some of the anaglyph images using their own skills and equally excellent software. I am also very grateful to Ken Herkenhoff and his colleagues at the U.S. Geological Survey in Flagstaff and Ken Edgett and his colleagues at MSSS for suggesting and generating the 3-D microscopic views, and to the many fellow Martian "travelers" at The Planetary Society, unmannedspaceflight.com, and elsewhere for suggesting some of their favorite photos (so many photos to remember!). While these folks have been amazingly helpful on this project, I've done enough meddling on my own to assert that any shortcomings in the selection, processing, or visualization of the images are completely my own, not theirs.

On a personal level, I thank Michael Bourret from Dystel & Goderich for believing in this project from the beginning and (gently) helping to get it to the launch pad, and Michael Fragnito, Melanie Madden, and Meredith Hale and their team at Sterling Publishing Co., Inc. for expertly guiding the initial edition, as well as this updated and expanded edition, to a safe landing. And as always, I am grateful to my family and friends for their support, patience, and indulgence while I was working to assemble this collection of wonderful Mars images and related stories.

JIM BELL
July 2013
Mesa, Arizona

PREFACE

I LOVED 3-D PICTURES as a kid. It was great fun putting on the silly red-and-blue glasses and then watching dinosaurs, bugs, and rockets jump out at you—like magic. One of my favorites was a 3-D ViewMaster cartridge that showed some of the early Mercury and Gemini astronauts and their spaceships—what a thrill! It felt like I was up in outer space with them.

For the past decade and a half, I've had the incredible privilege of being in charge of the color camera team for the Mars Exploration Rovers *Spirit* and *Opportunity*, the plucky, little wheeled robots that started driving around on Mars in early 2004 (as of this writing, *Opportunity* is still going strong!) and being the deputy in charge of even more capable color cameras on the Mars Science Laboratory rover *Curiosity*. Each rover carries pairs of wide angle and high resolution stereo cameras and microscopes capable of stereo imaging. When we designed the cameras and started thinking about how they'd be used on Mars, we all knew that taking 3-D pictures would be an important part of their daily work. However, to be honest, I didn't think that we'd get that many 3-D images— certainly not enough to fill a book. In fact, many of us didn't think that the rovers would land safely, or survive on Mars as long as they have. *Spirit* and *Opportunity* were designed by the creative and skilled engineers at NASA's Jet Propulsion Laboratory (JPL) in Pasadena, California, to last

OPPOSITE **Rocky Hollows**
Spirit acquired this 3-D Pancam mosaic while driving northeast across the volcanic plains toward the rim of Bonneville Crater. The impact that created Bonneville may have spread a large amount of previously buried material across the area.

for at least 90 Martian "sols" (a sol is a Martian day, about 24 hours and 39 minutes long), but they survived—and thrived—for much, much longer than that. *Curiosity's* mission has only recently begun, and it was designed by JPLers to last nearly 700 sols on Mars. The result has been an enormous data set of images, more than 300,000 so far, including thousands of glorious 3-D views of an alien landscape at many different scales of magnification and from a variety of perspectives.

This happy circumstance has provided me with the chance to combine my childhood love of 3-D pictures with my grown-up day job as a Mars rover scientist and photography specialist. The results are compiled here in what are, in my opinion, some of the "greatest hits" 3-D images from the rovers during their adventures on Mars. About ninety 3-D photos are shown here, along with (wherever possible) a natural color or false-color view of the same or similar scene for comparison. Many of the 3-D images are striking: Rocks seem to pop up off the page, and sand dunes change from gentle, wavy landforms to menacing driving obstacles. Perhaps most impressive is the way the topography of the landscape, which often seems flat when viewed in a normal photo, jumps out at the viewer when seen in 3-D. Seen in three dimensions, the rovers' wheeled maneuvers across rolling plains, up rugged, rocky ridges, and down slippery impact-crater slopes are even more impressive. These images also provide a sense of the challenges that my colleagues on the team have faced in navigating these remote-controlled robots across the terrain, and those faced by the rovers themselves, when they've had to switch to their own self-navigation software.

For me, the most difficult part of viewing these images is the almost complete lack of any gauge that could be used to indicate the size of objects. There are no trees, shrubs, empty soda cans, or even geologist's rock hammers in these photos to provide a sense of relative size or distance. And since the rover itself is the photographer, there's often no way to get a wheel or some other part of the rover into the scene to make up for the lack of scale or context. Many of the rocks and sandy ripples that are close to the rover are much smaller than they seem; conversely, many of the hills and ridges that are seen off in the distance are actually much farther away than they seem. I've tried to indicate some sense of size and scale in the captions, where possible, but the images are still somewhat visually misleading at times.

You'll notice a lot of interesting or colorful names of rocks, craters, hills, and other features throughout the text and captions: Wooly Patch, Pot of Gold, Home Plate, Purgatory, El Dorado, etc. Giving names to new places and things is something that explorers have been doing for a long time. Most of the names that we use are theme-related to the place under observation at the time we're exploring the area. For example, rocks and other features on and near a broad plateau that we called Home Plate in Gusev Crater have been named after famous players from baseball's Negro Leagues and the All-American Girls Professional Baseball League. Sometimes we were more methodical, as when we named craters in Meridiani Planum after famous ships of exploration *Victoria* was one of Magellan's ships; *Endurance* was the ship on Sir Ernest Shackleton's famous South Polar Expedition; and of course, *Eagle* was the ship that took Neil Armstrong and Buzz Aldrin to the Moon). Some of the names were suggested by school kids or members of the public who sent e-mails to various team members. Many times, though, names or themes just popped into teammates' heads on the spur of the moment—like rock-naming themes based on favorite ice-cream flavors, prairie wildflowers, or people and events from the French Revolution. None of these names are "official," though, in the sense of formally being recognized as feature names by NASA and the International Astronomical Union's planetary nomenclature committees. The only officially named features have been those of the major peaks in the Columbia Hills, named to honor the astronauts who lost their lives in the Columbia Space Shuttle tragedy in 2003, and three other prominent hills in Gusev Crater that are named in honor of the astronaut crew who lost their lives in the Apollo 1 fire in 1967. For most everywhere else, naming things is just a fun activity that helps us to remember the places we've visited.

The main object of this book is to showcase some of the best 3-D photos from the rovers. Some of the accompanying text describes the ways we acquired the images and how we used them every day for scientific studies and for making decisions about how to drive the rovers or position the instruments. I've taken the liberty of expanding the text beyond just describing the 3-D imaging, however, in order to tell other stories about these exciting missions. Many of these stories are the expanded versions of answers to common questions that I get when I give talks around the country about the rovers: How do you drive a remote-controlled car from across the solar system? Why didn't you install a wiper blade to clean the dust off the solar panels? How long are they going to keep going? Hopefully, these stories will help people understand some of the behind-the-scenes details about the kinds of decisions, skills, and experiences needed for a team of people to pull off an amazing feat like driving rovers across the surface of Mars. And hopefully, the subjects of the 3-D photos shown here will jump off the page to amaze and delight you, just like the 3-D photos I viewed as a kid delighted me.

ONE

Why 3-D?

SOMEWHERE ON THE OTHER SIDE of the solar system, a lonely traveler is making its way across an ancient, dry lakebed, taking pictures and making measurements, every day, to share with all of us back here on planet Earth. This is no ordinary traveler, though: it is a robot, designed and built by earthbound humans to work and thrive independently on another planet. And this is no ordinary planet: it is Mars, a world, it would seem, that used to be a lot like the Earth but that has somehow dramatically transformed into the cold, dry, barren place that we know today.

The traveler—travelers, actually—are the Mars rovers *Spirit*, *Opportunity*, and *Curiosity*. They are the hardy and intrepid robotic explorers that were launched from Earth in 2003, which landed on Mars in 2004, and that have been cruising around the Red Planet for more than four years, making one scientific discovery after another.

A team of literally thousands of people—engineers specializing in robotics, rockets, power and communications systems, and navigation; scientists specializing in geology, astronomy, chemistry, physics, and even biology; supporting managers and administrators—designed, built, tested, launched, and then remotely operated these rovers on Mars from Earth. The team has had one main goal in mind through what has been more than 15 years of planning and work on these missions: to use these amazing machines to project ourselves there in order to "virtually" explore Mars.

The rovers are extensions of ourselves and our very human capabilities. People can't yet go to Mars (though that day is coming), so we do the best we can given our limitations. The rovers were designed to be akin to mobile field geologists and, as such, possess many of the "senses" and abilities we would have if we could be out there exploring the place ourselves. We endowed them with mobility—the ability to move and drive and climb and descend. They move slowly, at a turtle-like pace, but with purpose and conviction, either relying on our judgment to help them navigate the remote terrain, or many times using their onboard software to make their own decisions about avoiding hazards or obstacles or otherwise finding the safest path to the next target of interest. The rovers have astonished us with their ability to scramble up rocky hills and descend down steep crater slopes, with *Spirit* driving more than 7 kilometers (4.5 miles) until its mission ended in 2010, *Opportunity* driving more than 37 kilometers (more than 23 miles) so far, and *Curiosity* more than 500 meters (0.3 miles) in its initial mission so far. They've turned out to be marvelously mobile Martian machines.

We gave the rovers the ability to "feel" their environment using agile, human-like arms that can reach out and touch the rocks and soils. The arms have "fingers," that can measure the details of the chemistry and mineralogy on Mars; brushing, grinding, or drilling devices which can clean off dusty surfaces or grind or drill small holes into rocks to allow their internal properties to be measured; and microscope-like high resolution cameras for taking close-up pictures at about the same resolution as a geologist's hand lens. If we were sending robotic geologists into the field, we figured, they should be equipped with hand lenses. After all, we would be.

The rovers' cameras, their "eyes," are the final, most critical piece of this extension of ourselves into the alien Martian landscape. To a geologist, having mobility is good, as is being able to take samples and detailed measurements. But the most important thing geologists do in the field is look because, while the land is telling a story, preserved in the rocks and soils, it is a difficult one to read. There may have been violent eruptions of lava or floods of water. Or maybe there were glaciers, or ancient, long-ago buried rivers or sand dunes. The story is

OPPOSITE **Egress**

This 3-D view of the *Spirit* rover rolling off its lander was generated by merging actual 3-D images of the lander on Mars with a 3-D computer model of the rover superimposed on the scene at the correct scale. The rover is about as tall as a typical ten-year-old kid.

there, in the geologic record, but deciphering the clues requires careful observations and a fair bit of detective work. Looking for clues in the land's features is the key to untangling its past. It's the same for robotic geologists on Mars, which is why we knew we would have to give the rovers the best pairs of eyes that had ever been flown into outer space.

Spirit and *Opportunity* have nine "eyes" and *Curiosity* has seventeen! The microscopes are on the rover's arm; fisheye-lens cameras are near the ground on the front and on the back (these are the hazard avoidance cameras, or Hazcams); wide-angle cameras are at the top of the mast (the navigation cameras or Navcams); and high-resolution color panoramic cameras (Pancams and Mastcams) are also placed at the top of the mast for scientific observations. *Curiosity* has an additional mast-mounted high-res camera called the Remote Micro-Imager (RMI). Most of the rover cameras take black-and-white photos; only the Pancams on *Spirit* and *Opportunity* and the Mastcams and Mars Hand Lens Imager (MAHLI) on *Curiosity* can take color pictures, by using thumbnail-sized filters to shoot, for example, red-, green-, and blue-filtered images that we can then combine later (once the photos are radioed back to Earth) into full-color (RGB) images. The resolution of the color cameras ranges from about the same as that of a typical pair of human eyes on a person with 20/20 vision, up to three times sharper. However, some of the color filters give the rovers superhuman color imaging capabilities by enabling us to see in the ultraviolet and infrared part of the spectrum through their eyes.

Having pairs of eyes on a camera system enables the world to be viewed from two perspectives. The resulting parallax—just like the one that exists for our own eyes—is processed by the rover's computer to yield a stereo view of distance and depth. Even images from the one-eyed handl lens microscopes on the rover's arms can be used to "see" in stereo, by moving the arm to different positions in order to view the scene from a variety of angles. We can take the images viewed from the left and right sides and process them in a computer on Earth to simulate what the rover saw on Mars. The result, as you'll see in this book, is eye-popping. Rocks "jump" out of the photos. Sand dunes take on a feeling of waves cresting on the beach. Hills, ridges, and valleys suddenly appear where before there was only a flat world of two dimensions.

The Mars rover stereo images (also called "anaglyphs"—from the Greek word for "carved in low relief") can be stunning, even disorienting, to view, given their literally otherworldly nature. They are more than just interesting 3-D photos, though. During the mission, we used stereo images like these every day for practical, pragmatic purposes. For example, if we wanted to measure the size of a rock or a sand dune for scientific analysis, or assess the slopes and headings of different rock layers in the course of generating a geologic map, or accurately calculate the distance to a certain hill or ridge to triangulate our location, we'd use the parallax of the stereo images and the known resolution of the camera as a distance- and slope-finder. Or when we were planning to drill into a specific rock, first we'd take stereo pictures of the surface of the rock in order to build a 3-D computer model, and then the engineers could use that model to choose the specific location and "angle of attack" for the grinding or drilling.

Every day, the rover drivers (believe it or not, there is a small team of "licensed" Mars rover drivers who work at the rover's mission operations center at JPL) use stereo images to build, inside a computer, a virtual-reality 3-D model of the terrain around the rover. They can then practice "driving" a computer model rover in this computer-generated facsimile of the real Martian world. The rover drivers have always worked closely with the scientists on the mission and the computerized rover/topography models to make sure that it was possible (and safe) to drive the rover to the places the scientists wanted to go. If things got dicey, they would then have to either find a way around obvious obstacles or over less steep slopes, or work with the scientists to choose a different target destination. Once the path forward looked good, the drivers would build the final "drive sequence" and program the real rover on Mars to carry out the same driving instructions that had been used to propel the computerized rover on Earth. Essentially, you can think of the rover's stereo images as part of an elaborate interplanetary video game being played every day.

We see and experience Mars through the eyes of the rovers. Spending time exploring the 3-D images taken by them is my favorite reward for having helped endow them with human-like vision. Our gift to them was vision and independence; their gift to us has been to show us the depth, texture, color, and beauty of the Mars many of us wish we could see one day with our very own eyes.

TWO

Machine Vision

THE ROVER'S STEREO CAMERAS and computer are modeled after the human eye and brain, which come equipped with one of the most incredible stereo imaging systems available. A pair of human eyes is a color, binocular stereo "camera," if you will, connected in a real-time feedback loop to a phenomenal central processing unit (our brain) that rapidly scans images, recognizes patterns, and performs necessary routine mechanical maintenance. Of course, our eyes and brains have evolved to be efficient at these tasks, many of which can be critical to survival (for example, Tiger approaching! Run!). The resulting combination of two high-resolution imaging "cameras" providing 3-D input to our superfast decision-making internal "computer" is the envy of and model for engineers, who then try to mimic these capabilities in robots and other electromechanical imaging systems. These researchers work in what is sometimes referred to as the "machine vision" field. There are strong links between this community and scientists and engineers working in the fields of computer programming, bioengineering, artificial intelligence, and, of course, space exploration.

Machine vision is just what it sounds like: the attempt to give machines the ability to "see" in a human sense and to make intelligent decisions based on what their sense of sight tells them. The potential of practical applications for successfully enabling machines to see and decide things on their own is enormous. One area in which this field has already taken off is manufacturing and process control. As products fly past on high-speed conveyor belts, for example, machines have been programmed with the ability to search for anomalies or defects (Is that top the right color? Is the package dented?) at a rate much faster than a human being could, and with fewer errors. There are a huge number of other applications for machine vision, like medical research and diagnosis, security and surveillance, transportation (navigation, traffic monitoring, hazard avoidance), and remote sensing. Machines are reacting to what they observe all around us, and for the most part our society now takes their help for granted.

When those of us on the Mars rover design team started thinking about the kinds of capabilities the rovers would eventually be endowed with, machine vision was a critical part of the equation—especially for driving. It had to be. In essence, the rovers' sight had to "evolve" to be as efficient and useful as possible to their survival in the Martian environment. Perhaps the most environmentally relevant factor was that the rovers would be anywhere from about 60 million to more than 300 million kilometers (35 to 186 million miles) from their controllers back on Earth. Even at the speed of light, that meant radio signals carrying pictures and other data could take from about 4 to 17 minutes just to get from Mars to Earth, and it would take just as long or longer for instructions from home to get radioed back to Mars. By the time a hypothetical cry for help could be received from one of the rovers ("Hey, those commands you sent me are going to cause me to drive over the edge of that cliff. Should I stop?"), it might already be too late to send a reply. The alternative would be to send commands to make the rover do some minimal, safe activity and then to wait for images and other data to confirm that activity was not hazardous before going on to the next one. However, this method could be inefficient or impractical for some kinds of terrains or driving circumstances because each of these mini-decision cycles might take an entire day to complete (we're not able to be in constant communication with the rovers because NASA's Deep Space Network communications antennas need to support many other space missions, plus the rovers are out of view of the Earth for about half of each Martian day). Clearly, the rovers had to be able to make their own decisions sometimes, based on what they saw. And

OPPOSITE Desolate Ridge
This 3-D view was taken by *Spirit's* Pancam instrument. The goal was to get the rover to the horizontal ridge, seen halfway from the top in this view; stereo images like this one provided the team with the critical information needed to safely plan the drive.

those decisions had to be the right ones, given that about $8400 million of taxpayer funding had been invested in Spirit and Opportunity and $2.5 billion in Curiosity to land them safely on Mars!

While it may be essential in some cases, it's still not easy for scientists to turn over control of critical driving or data-gathering decisions to a robot running a computer program. The rovers are not "intelligent" **per se**; they are only as smart as their software, which in turn is only as good as its programmers. Even though their software was built and programmed by teams of very clever people, and was tested extensively, it can still be a difficult thing for people to be out of the loop, and trusting a robot to perform well in a hostile environment so far from home. For the most part (after a bit of on-the-job training and a few software tweaks), the rovers and their capable software designers have earned the trust of the entire science and engineering team. The truth is that we never could have traveled so far and sampled the geology, chemistry, and mineralogy of so many different kinds of places that we've visited if the rovers hadn't been provided with such great stereo cameras, coupled with autonomous decision-making capabilities.

I worry a little about the effect this positive experience might have on future missions and experiments, though, and whether it might place us on some kind of a scientifically slippery slope. How does a machine decide whether or not a picture contains interesting or uninteresting information? What factors go into choosing among multiple targets for more close-up examination? What kinds of measurements get priority if resources are tight? These kinds of questions have been debated extensively among our team and among people running space missions in the past, and the answers can't always be boiled down to a simple algorithm, especially for unanticipated discoveries or for more "qualitative" or context-dependent areas of research like geology. Sometimes, even in space exploration, people use their intuition—that is, they act on a hunch or roll the dice to see what comes up. How do we program those kinds of very human traits and propensities into a machine? Is there really such a thing as artificial intelligence?

These kinds of issues are actually part of the broader "humans versus robots" debate that has been going on for decades, and which is still going on right now within the space program. Which is better: a long-term, continuous program of relatively inexpensive but technologically and intellectually limited robotic missions, or more expensive and risky but less frequent planetary exploration conducted by humans? It's a tough call, and there are camps with firm opinions entrenched on either side of the debate.

It is easy to imagine the side of the debate on which most scientists who are heavily involved in robotic exploration missions like the Mars rovers come down on. However, my personal opinion on the "humans versus robots" question may be a bit unexpected: I'm a big supporter of human exploration. I believe that without a strong, vibrant, and destination-driven program of human space exploration, future successful robotic missions like the ones employing the rovers just aren't going to have the momentum to go forward. It was the *Apollo* and *Skylab* astronauts, after all, who first got me interested in and excited about space exploration and a possible career in planetary science. I think that kind of motivation is necessary at the national level as well. It is the human adventure of astronauts in space (including its dangerous, perilous, and tragic aspects) that excites and inspires the public and that helps to galvanize support in Congress and the Administration for continued space exploration funding—whether human or robotic.

My view is framed by more than just intangibles like inspiration, education, and national prestige, though. I think that people are going to be needed on other planets to use their judgment, experience, and understanding of context to truly promote new scientific discoveries. Robots have made possible a huge number of discoveries in our solar system; however, many of them occurred because it was the first time we had flown past, orbited, or landed on a particular place, so our robotic emissaries were (to borrow a phrase) "picking the low-hanging fruit" from our species' first phase of solar system exploration. We'll be finishing up that phase over the next several decades, however, which means that new discoveries are going to be more challenging to make:

The remaining fruit is higher up on the tree. Robots are getting smarter all the time, but it's not clear whether robots' brains will ever be able to think on the fly or react to a hunch the way a human brain can. People— that is, astronauts with experience—may be required to tease out the most important discoveries of the future.

Sadly, the Space Shuttle and the International Space Station missions don't seem to have inspired or motivated the public the way the *Apollo* Moon missions of the late 1960s and early 1970s did. I think this makes the present an extremely risky and uncertain time for the future of space exploration. If NASA and other agencies can't get people—teachers, kids, space enthusiasts, and the general public—interested in space exploration, the entire enterprise (human exploration, robotic missions, and basic research) could wither on the vine. Part of the reason the Mars rovers have been so popular in the media and among the general public is precisely because they seem to offer the closest thing yet to having people on Mars. We can all move, virtually, from place to place there. We see Mars through their machine vision as we would see it through our own eyes. It's a destination we can imagine visiting, and there appears to be a nascent thirst among people to voyage there. Will we muster the will, the know-how, and the courage to make the trip ourselves? I hope so. The future of both human and robotic space exploration may depend on it.

THREE

Can't Kill 'Em

I VIVIDLY REMEMBER SITTING in a room listening to engineers talk about how the Mars rovers, *Spirit* and *Opportunity*, would die. The setting was an engineering presentation at JPL early in the history of the project. The room was full of scientists and engineers and managers, many of whom I'd never met but most of whom would go on to play critical roles in the success of the rover missions. The rovers had not yet been dubbed *Spirit* and *Opportunity*, just MER-A and MER-B, respectively. They weren't even born yet; in fact, they were still just twinkles in their designers' eyes. Still, these people were already trying to figure out what would eventually kill their creations.

Cold, dry, barren, dusty, distant, and unforgiving: These words describe the environment the rovers would eventually be expected to call home. As scientists studying Mars—the most earthlike of all the other worlds in our solar system—we are sometimes guilty of glorifying what is, in reality, a truly desolate and inhospitable place. The pictures may at first look familiar, like a sunny desert scene, but with nary a cactus in sight. One expects to see a tumbleweed now and then, or a bleached jackrabbit skull. But Mars is nothing like a desert on the Earth. Far from it: The average temperature on the Red Planet is −50° to −60°C (−58° to −76°F)—close to the coldest temperatures ever recorded on Earth. On a balmy summer day in the Martian tropics, it might get up to 5° to 10°C (41° to 50°F) for a short while, but during a winter night near the poles, it drops down to −140°C (−220°F), an inconceivable cold during which the air itself (almost all of which is CO_2) freezes onto the ground. Even on the driest deserts of the Earth it rains, though perhaps only centimeters per century. As far as we can tell, it hasn't rained on Mars in perhaps 2 **billion** years or more. Talk about a drought!

So the rovers had to be built to survive the cold, with insulation and heaters to keep sensitive electronics warm, and batteries and solar panels to provide enough electricity to keep those heaters running at night. In fact, I've come to recognize that the single most important thing that the rovers do on Mars every day is recharge their batteries using electricity from the solar panels. If the batteries were to become drained so that we couldn't run the rovers' internal heaters at night, the temperature of the electronics, circuit boards, and other internal components would get so low that something would undoubtedly break. And that would be the end of the mission.

There were plenty of other ways the rovers could have died on Mars, or even before they got there. They could have blown up on the launchpad in Florida if the Air Force/Boeing Delta II rockets that were used to launch them failed. They could have been shaken to pieces by the forces and vibrations of the launch itself. There could have been a problem with the upper stage of the rocket before the rovers left the Earth's orbit. Or a problem could have arisen with their cruise-stage power supply or other electronics during their seven-month flight to the Red Planet. Solar flares or galactic cosmic rays could have zapped the computers during the cruise as well. Indeed, that almost happened, as some of the largest solar flares in recorded history erupted during *Spirit* and *Opportunity's* trips to Mars in the Fall of 2003, bombarding both vehicles with high-energy radiation. Luckily, even this potential cause of death had been anticipated to a degree, and the built-in shielding to protect the rover computers and other electronics proved robust.

Then, of course, the rovers could have burned up in the Martian atmosphere if the entry trajectories were off by just a bit, or they could have crashed to the surface, shattering into a million pieces, if the parachutes failed to open or if any one of dozens of pyrotechnic devices failed to fire at just the right moment in their Rube Goldberg–style, airbag-assisted landing systems. Or they could have landed safely, only to tumble to

OPPOSITE **Escaping Eagle**
Opportunity spent nearly two months studying Eagle Crater. This Navcam stereo mosaic was shot after the rover struggled to drive out of the crater. The rover finally escaped and was able to take this look back at its tracks leading back to its former Martian home.

their deaths off their lander platforms. What a catastrophe that would have been—to come all that way and then to trip over the finish line and collapse! It was a failure scenario that someone was contemplating, though, which meant that there needed to be a plan to deal with it, and a backup plan in case that one didn't work. Even if we had landed them safely on the surface, with six wheels in the dirt, some unlikely software glitch or errant command sent from Earth could have shut off the transmitter or fouled up the software and effectively killed a rover. Indeed, this would have happened on *Spirit* just a few weeks after landing if not for some incredibly prudent contingency planning by the rover engineers and software designers. Even the dryness of the Martian air could have killed them, potentially creating a lethal dose of static electricity as the rovers drove around. Care had to be taken to design circuits that were well-grounded and systems that were not prone to static discharges.

If the cold or the dry didn't kill them, then maybe the dust would. Martian dust is famous—it gives the planet its characteristic reddish color, and astronomers using telescopes have observed it for centuries, on occasion completely covering the planet for weeks on end. Mars dust is extremely tiny stuff. It's basically composed of smoke-sized, magnetic, and slightly rusted mineral particles only a few thousandths of a millimeter in size. Therefore, like soot and smoke on the Earth, it gets into everything. Wheels, gears, electronics, camera lenses—everything—will eventually get dusty and dirty, which could cause something to wear out or to jam or gum up. Most frightening was the threat posed by dust in the air (lifted by dust devils and dust storms) always coating everything—like the rovers' solar panels.

Perhaps the favorite potential cause of rover death, often discussed, was the slow choking off of their electrical supply by dust settling onto the solar panels and slowly, inexorably, blocking out the sunlight. In fact,

OPPOSITE **Bump in the Road**

When the rovers are using their onboard software to "drive themselves," Hazcam images provide the information needed for the rover's automated onboard obstacle-avoidance software. This photo shows the front wheels driving backward over a 5 cm (2 inch)–high sand dune.

Viking Orbiter view of clouds around Olympus Mons, the highest volcano in the solar system.

during the first few months after the 2004 landings, we could see this happening in the daily power levels the rovers reported back to us. We were seeing two effects, actually. First, both *Spirit* and *Opportunity* landed in the midst of the southern hemisphere summer, and as summer made its transition into autumn, the Sun was getting lower in the sky every day, causing our power levels to steadily decline (the solar panels generate more electricity when the Sun shines directly on them, instead of at an angle). We had planned for that power-level drop, and we knew that if the rovers could survive until the winter solstice, the Sun would start getting higher in the sky again and their power levels would go back up.

The other effect we saw in our daily power levels was a gradual decrease in power as the solar panels got ever dustier. We could see this in the pictures, too: Our once shiny, gleaming blue rover deck was becoming a dusty, faded brownish red. The power kept dropping as the panels got dustier and the Sun got lower. Many of us could see the end coming, but no one really wanted to talk about it. We just kept pressing on, trying to get as much scientific exploring done with whatever time these amazing space vehicles had left to them.

And then, one day while *Spirit* was climbing to the summit of Husband Hill (the tallest peak in the Columbia Hills complex near the middle of Gusev Crater), a dust devil or a rogue gust of wind came along and—**swoosh**—cleaned most of the dust off the solar panels! All of a sudden our power levels jumped up. The deck was once again shiny blue, and we suddenly found ourselves with the power (literally) to do much more science and remote driving than we could do before. This happened multiple times, and on *Opportunity* too, over the course of the mission. Certainly there was no way we could have anticipated the winds on Mars helping to keep the solar panels clean! As in life, sometimes in the space exploration business, you get lucky. For *Curiosity*, we avoided this problem by making the rover nuclear powered (like the *Viking*, *Voyager*, and *Cassini* missions) rather than solar powered.

People often ask me why we didn't design some kind of wiper blade, compressed air sprayer, or other gadget to keep dust off the *Spirit* and *Opportunity* solar panels. It's a great question, and the answer may not be at all obvious. I tell them that it would have been relatively easy for the engineers to design something to clean the panels every so often—it's just a mechanical problem, after all, not rocket science. However (and here's the catch), there's only a fixed amount of money, mass, volume, power, and time available when you're designing, building, launching, and then operating a space mission. So adding some kind of panel wiper system would have meant getting rid of something else to offset the resources this new system would need. What would we remove? A camera? A spectrometer? A redundant radio transmitter? An extra battery heater? It's a tough choice because it's a zero-sum game. Our engineer friends assured us that the solar panels would give us the electricity we needed to keep the rovers going for at least 90 sols on Mars, even if they got dusty. So we decided to risk it and not do without any of the instruments or other systems in exchange. And they were right. The fact that we went beyond 90 sols—way, way beyond—was a fortunate bonus and a by-product of great design, careful planning, and yes, plain old dumb luck.

With spacecraft and space mission instruments, it is often said that things either break on their first day of use in their harsh new environment or they survive essentially forever. *Spirit* and *Opportunity* and now *Curiosity* are testaments to that aphorism, and to the people at JPL and other government labs, universities, and hundreds of vendor companies around the world. They designed these amazing machines, provided parts and advice, wrote the software to run them, and tenderly guided them across the rocky, sandy plains far from their former home. Their designers came up with dozens of ways to kill the rovers, but it took Mars more than six years to kill *Spirit* (which succombed to dusty solar panels and stuck wheels in the soft sand), and *Opportunity* and *Curiosity* are still going strong. I think these rovers are like wild animals released into their native habitat, thriving in the environment they were built for, and truly fulfilling their birthright.

FOUR

Daily Grind

SOME DAYS I FEEL LIKE A TOURIST on a vicarious but virtual vacation to a far-off land. Those of us involved in the day-to-day operations of the Mars rovers come in to work in the morning and fill our computer screens with the latest pictures and other data from the Red Planet. Sand dunes, wispy clouds, frothy volcanic rocks, wheel tracks, and sometimes strange and alien landforms and features greet us daily. Many times it's the 3-D images, like the ones showcased here, that grab our attention the most. The strange part (as if our jobs aren't strange enough already!) is that many of us have gotten used to it. In fact, I don't really even remember what my day job was like before the rovers got to Mars. (What did I do with all that free time?) We've come closer than any other people ever have to **being** on Mars and experiencing the place as if we were really there.

When the rovers frist landed on Mars, all of us earthbound scientists, engineers, and rover drivers had to live our lives on "Mars time." That is, we had to react to and operate by the cycle of sunrises and sunsets that the rovers experience on Mars, because sunlight is what provides the heat and power that the rovers need to work most effeciently (even nuclear-powered *Curiosity*). However, a Mars day is about forty minutes longer than an Earth day, so our workdays here on Earth—and our lives in general—were slowly shifting by an increase of forty minutes each day, because the Sun rose on each rover forty minutes later each Earth day. It was fun for a while, working crazy hours and feeling detached, in a way, from our home planet even though we'd never actually left. I think it helped build a bond between us and our robotic friends on Mars. "Who's controlling whom here?" we wondered, while trudging in to start our shifts at 3:00 A.M.

During those first three to four months of the *Spirit* and *Opportunity* missions, several hundred of us lived and breathed Mars and rovers at JPL in Pasadena. The rovers had landed near the Martian equator (lots of sunlight) but on opposite sides of the planet: *Spirit* in Gusev Crater, a possible ancient lakebed the size of Connecticut, and *Opportunity* in Meridiani Planum, a flat, sandy plain the size of the western United States where evidence of water-formed minerals had been detected from orbit. As the Sun set on one of the rovers, the rover would radio its latest pictures and other telemetry information to us and then proceed to "sleep" for the night, hunkering down with its heaters to keep its systems warm and alive until the next morning. At the same time, on the other side of the planet, the Sun was just rising on the other rover, "waking it up" to begin its workday.

Our attention first focused on looking at the pictures and other measurements just sent down to Earth by whichever rover was then going to sleep. The clock was ticking on us all the time: We had to have a new set of instructions ready to transmit to that rover as soon as it woke up the next morning. That meant that we had to process the images and do some very quick analysis and assessment work to help make informed decisions about what to do, usually within just a few hours of when the data arrived. Should we tell the rover to drive? If so, which way? Or should we put the arm down on some targets to make detailed measurements of the chemistry and minerals, or photograph some big panoramas? While we always had well-thought-out strategic goals that we were aiming to meet, every sol presented us with new tactical challenges and opportunities to address. It was grueling but exciting work.

Once the constraints on the next sol's activities were understood (power, data volume, etc.) and the high-level tactical scientific decisions had been made, the part of the science and engineering team that concentrates on downlink—the assessment and analysis of data sent "down" to us from Mars—had essentially finished the most stressful part of its workday. Sometimes there would be time to look

OPPOSITE **Mazatzal**

This is a colorized *Spirit* 3-D Navcam mosaic showing the area near the rock Mazatzal. About 2 meters (6.5 feet) across, Mazatzal is the bright rock on the left of this view. A Rock Abrasion Tool (RAT) hole ground into the rock revealed it to be made of dark volcanic minerals.

at the data more in depth, or to more thoroughly investigate some scientific or rover system mystery that appeared in the data, before the next downlink would arrive. Sometimes, though, the downlink assessment and tactical planning process would have been so lengthy and stressful that once over, all you would have wanted to do was lie down on a cot and sleep.

The action would then continue on to the "uplink" process—the act of building the detailed instructions and commands to send up to the rovers on Mars. The clock was still ticking—we might have eight hours or so left until the rover was going to wake up. The uplink team was generally made up of different people from those on the downlink team. This was done partly for practical reasons, such as trying to keep people's workday to shifts under twelve hours. But partly it also reflected the fact that sending commands to run rovers and other space instruments is often much more of an engineering and software detail-oriented activity than is science or telemetry downlink assessment. People working on the uplink shift listen to the downlink reports and follow the tactical decision-making process; then they have to be the ones to implement the results of those decisions—in detail. It's one thing to agree that you're going to take a panorama but another thing entirely to actually write the computer commands that point the camera, turn the filter wheels, take the exposures, etc., and then string together all the commands into detailed sequences that will be radioed to Mars.

Uplink people tend to be detail-oriented—careful, methodical, double-checker types. They are the kind of people who can spot a typo within seconds on a page full of words. They also tend to be extremely creative problem-solvers, and many of them are excellent at visualizing the situations they're writing sequences for and doing the 3-D math (often in their heads) needed to understand the detailed pointing of the instruments. They are exactly the kind of people you **want** to have driving a rover or an instrument like a camera, because they rarely make mistakes. I'll confess that I'm not an uplink person— I don't have the patience for the level of detail and scrutiny their job requires, and I can't do 3-D vector math in my head (or on paper, for that matter). I'm much happier working with the pictures and other data on the downlink end of the process. It's all a matter of personal preference and individual skill and training. There is variety enough in the kinds of jobs required to be done to make for a team of people with extremely diverse interests and capabilities.

The uplink process results in the creation of a long list of sequences that the rover will start to run once it wakes up in a few hours. The list is made to fill up every available minute of time and every available watt of power. It's essentially a "to do" list for the rover, with a bunch of time tags telling it when to do each activity. For example, "10:53 A.M.: Begin battery recharging cycle; 11:20 A.M.: Begin drive 10 meters forward; 12:00 P.M.: turn 20 degrees left; 12:50 P.M.: drive 10 more meters forward; 1:30 P.M.: acquire 360 degree panorama," etc. The list is laboriously assembled, sequence by sequence, and is then checked and double-checked in the remaining hours before being transmitted. Sometimes the team will need to try out sequences on the test rover or simulator at JPL before sending them to the real rover on Mars. Once everyone's happy with the list, it's converted into the ones and zeros of spacecraft radio signals and transmitted by one of the NASA Deep Space Network telescopes, directly to the newly awakened rover. Then, we're out of the loop—the rover goes off and does everything on the list autonomously, and we usually don't hear back from it until the end of its day, when it radios back the results from that day's activities.

Meanwhile, around the same time the list described above is finally radioed up to the rover, the rover on the other side of the planet (remember, there are two) would have just finished working through the list that we sent it yesterday, radioed its results back to Earth, and be getting ready to sleep. A second subset of the rover team would then begin working through the downlink assessment, tactical planning process, and finally the uplink sequencing process for that rover, meeting the deadline of getting its new list uplinked before it woke up the next morning. When that rover would wake up and get its new to-do list, the other rover would be radioing back its results, in keeping with its daily agenda, and so on. One of them was always working during the daytime, and we were always working on planning the next sol for whichever one was currently sleeping. Essentially, the Sun never set on our little Martian empire! The only ones getting any decent sleep were the rovers....

The 3-D images taken by the rovers were a critical part of this decision-making process every sol. Drives could occur only if the rover operators understood the topography and potential obstacles the rovers might face and felt confident that the rovers could negotiate parts of the terrain on their own, if necessary. The arm could be positioned and its instruments deployed onto rocks or other targets only if the shapes of those targets and the nature of the "work volume" where the arm could move was fully understood. Many times, we were basing our choice of scientific targets on detailed 3-D stratigraphy and on the geologic relationships between layers of rocks or other features that we could assess using the stereo vision capabilities of the cameras.

Working on this crazy cycle of round-the-clock care and feeding of robots on another planet was certainly a unique experience, but shifting our jobs 40 minutes later each day turned out to be so foreign to our natural circadian rhythms that it began taking a toll on our bodies and minds. By the end of the rovers' 90-sol primary mission on Mars, people were tired, irritable, and yearning for the good ol' 23- hour 56-minute day that we were used to back on Earth. So the mission managers brought us all back to our home planet, sent most of the scientists back to their home universities and labs, and switched us to an "Earth time" schedule for the rest of the mission. It was a little sad to separate ourselves from the daily cycles of the rovers and from the nerve center of the operations at JPL, but there really wasn't any other choice. It's just not possible to live on two planets at once, especially if you want to have a real life on one of them.

OPPOSITE **The West Spur**

This Pancam 3-D mosaic shows the slopes and ridges of the West Spur of the Columbia Hills, a region *Spirit* arrived at after nearly five months and more than 3.2 kilometers (2 miles) of driving.

The rover missions have since been run on Earth time instead of Mars time, and the team, now distributed around the world and networked through telephones and the Internet, has become extremely efficient and streamlined. We still go through the daily downlink and uplink cycles, but the decision-making and sequence-making processes generally take much less time than they did before. To save money (and perhaps some marriages), the managers eventually decided to give the team weekends off. However, the rovers don't get a break: On Fridays we plan each rover's activities for Saturday, Sunday, and Monday. The mismatch between Earth time and Mars time also means that occasionally we have to "skip" an Earth day planning cycle and instead plan two Mars days in a row. It sounds complicated, but we've actually gotten into quite a workable groove. Running the mission on Earth time and using so-called "distributed operations" has turned out to be a successful and efficient way to explore Mars remotely, and it has allowed team members to share more deeply the amazing results from these rovers with their families, neighbors, students, and broader local communities around the world.

FIVE

Not Your Grandfather's Mars

THE DISCOVERIES ENABLED by *Spirit*, *Opportunity* and *Curiosity* have fundamentally changed our view of Mars. The way that science operates best is by people making observations and developing models of those observations in order to pose hypotheses that can be tested. In the history of Mars exploration, several decades of previous spacecraft observations had revealed vast channels and valley networks carved into the rock, huge past rates of erosion, deposits of possibly water-formed minerals, and even some geologic evidence for rainfall, lakes, and glaciers. These observations led to the hypothesis that Mars used to have liquid water on its surface, perhaps for a significant span of time early in the planet's history. The idea led to the further hypothesis that such a warmer and wetter Mars may have been a **habitable** and much more earthlike world. These hypotheses are testable, so verifying them by studying the geology, geochemistry, and mineralogy of specific regions on the surface was the primary goal of the rovers and their instrument suites.

One of the most recent profound discoveries about life on our own planet is that it is dominated by simple, single-celled organisms, and that these organisms (with which all more complex organisms ultimately share common ancestors) occupy and thrive in an incredibly wide range of niches and environments. From acidic pools to arid deserts, from the bottom of the seafloor to the tops of the world's coldest glacial peaks, life abounds on Earth.

What makes a world habitable? This has become a central question for astronomers, planetary scientists, and biologists studying Mars and other heavenly bodies, as well as for geologists, biologists, paleontologists, and planetary scientists studying the Earth itself. Life is found essentially anywhere on Earth where it can avail itself of three main ingredients: water, energy, and organic molecules. Earth is a water world, three-fourths covered by salty oceans, harboring a global network of surface and subsurface freshwater, and still actively leaking out water and other volatile gases from the deep interior through volcanoes. Our planet has been described as being in a perfect "Goldilocks" place for water in our solar system: Not too hot (close to the Sun) to allow the water to boil off as on Venus, and not too cold (far from the Sun) to keep the water in a deep freeze like on Mars today. In other words, just right.

The Sun provides the main source of energy that enables life on our planet, through almost magical processes like photosynthesis, which converts sunlight into energy that can be used by cells (photosynthesis is to our cells what the solar panels are to the rovers). There are other ways to harvest energy, however. For example, bacteria deep underground extract energy from the rocks by slowly oxidizing them. Exotic life forms at deep mid-ocean ridges extract energy from volcanic heat and sulfur-rich gases and mineral deposits. Sunlight is an abundant energy source, but it's not the only source required for life—especially life underground or underwater—to thrive.

Earth is also endowed with a rich supply of organic molecules—the building blocks of life as we know it (there have been serious studies about "life as we don't know it," but that is extremely hard to search for and identify...). The elements carbon, hydrogen, nitrogen, oxygen, phosphorus, and sulfur can combine in ways that are simple or phenomenally complex to form the familiar organic molecules like proteins, carbohydrates, amino acids, and ultimately RNA and DNA, which constitute the basis of life on Earth. The Earth was formed with some of these chemical elements "built in," but others have been "delivered" to the Earth by comet and asteroid impacts throughout time.

OPPOSITE **Wishstone**
This *Spirit* Microscopic Imager 3-D mosaic shows part of a RAT hole ground into the rock called Wishstone. The rock's inner surface shows that it is composed of fragments of many different sizes and textures, possibly indicating that the fragments were fused together by some violent geologic event.

Might other planets, Mars in particular, have had or still have these three main ingredients? And might they exist or have existed in the right proportions to enable these places to be habitable? Mars has ample sunlight available (though only about half as much per square meter as the Earth). It also appears to have an inventory of rocks and minerals that could provide alternative energy sources similar to the ones exploited by organisms living deep inside the Earth.

Does Mars have water? Absolutely—we've known for decades that there is water ice in the polar caps and a tiny amount of water vapor in the atmosphere. Recent space missions have also discovered large deposits of underground ice at high to middle northern and southern latitudes. However, life (again, as we know it) appears to require *liquid* water, as a medium for chemical reactions and as a way to enable the mobility needed to bring in nutrients and to take out waste. So, does Mars have liquid water? Not today, at least not on the surface. The temperatures and pressures are too low for liquid water to be stable at or very near the surface (it would quickly "boil" off into water vapor). However, evidence from orbital data and now from rover data acquired at three specific sites indicates that Mars did have liquid water on and/or very near its surface long ago, when the planet was young. Minerals have been found that can only be formed in liquid water; some of them like clays and sulfates still have the H_2O or OH molecules derived from that water in their mineral structures. Geologic features, like preserved shallow water-wave ripples and the iron-rich spherical grains known as "blueberries" at the *Opportunity* rover site (which appear to have precipitated out of groundwater like stalactites or stalagmites in a cave), or the rounded streambed-like pebbles and conglomerates at the *Curiosity* rover site, also provide evidence of water at one time flowing over the surface and saturating the rocks. The evidence seems incontrovertible: Mars used to be warmer and wetter. Its environment used to be a lot more like the Earth's.

OPPOSITE New Kid in Town

In May of 2011, the Mars Science Laboratory rover *Curiosity* began a series of final driving and other tests at the Spacecraft Assembly Facility at NASA's Jet Propulsion Laboratory. The rover was shipped to the Kennedy Space Center in June 2011, and then launched to Mars on November 26, 2011.

Does Mars have organic molecules? This is a more difficult question to answer. Experiments on the Viking Lander missions in the 1970s were designed to answer this question, and the results are widely interpreted to have been resoundingly negative. However, those missions literally only scratched the surface, and the landing sites didn't turn out to be particularly interesting or perhaps even very "water-related," compared with the rover sites or many other potential landing sites on Mars. Like Earth, throughout its history Mars has been bombarded by organic compound–bearing comets and asteroids, and Mars was almost certainly endowed with a built-in inventory of organic molecules, or at least their precursor elements. So the answer is potentially, "yes," though we lack the slam-dunk proof that we have for other parts of the habitability question. Indeed, this is exactly why Mars Science Laboratory *Curiosity* rover carryied an elaborate organic chemistry instrument to its landing site. *Curiosity* is within an ancient elaborate impact crater named Gale, a place chosen for its geologic and mineralogical evidence of past surface water. The rover's initial organic chemistry experiments on the materials exposed there have not revealed unambiguous evidence for Martian organic molecules, but it's still early in the mission and the most interesting places in Gale seen from orbit haven't yet been visited and studied in detail. *Curiosity's* organic chemistry results over the next few years could be extremely enlightening and could help fill an important gap in our understanding of the habitability of Mars.

The rovers have helped us to make a giant leap forward in our view of the Red Planet. To my grandparents' generation, Mars was just a reddish point of light in the sky and the topic of often wild, sensational speculations about its possibly nefarious inhabitants. It's not our grandparents' Mars anymore, however. We've been there and have traversed (albeit virtually) its ruddy surface. No longer do we have to speculate about whether Mars has potentially habitable, earthlike environments—we now know that it once did. We searched for the "smoking gun"—evidence of the planet's watery past—and we found that evidence, either lurking there in the rocks and minerals or popping out at us as soon as we looked up close, much like 3-D images on a printed page.

The rovers have provided new and important information about exciting, previously habitable places on Mars, but by working with instruments on many different orbiters high overhead, the story's scope can be expanded to include the entire planet. What other environments there might be habitable? Where should we send the next Mars missions to maximize the chances of answering that question? There's also a burning question that the rovers haven't been able to address: How long was Mars earthlike? If it was a mere flash-in-the-pan early in the planet's history, then there may not have been enough time for life forms to gain a foothold. However, if Mars remained earthlike for an era of "real" geologic time—a billion years, maybe more—then we might be able to make a case here in our own backyard for parallel planetary development.

How exciting and meaningful it is to realize that we're actively trying to answer one of the most fundamental, far-reaching questions we can ask as human beings: Are we alone? Even if we were to discover only simple forms of past or present life on another planet such as Mars, the implications would still be profound. It would mean that there might be other habitable environments right here in our own solar system, and that more broadly, the universe is potentially teeming with life. Maybe one day we will be able to reach across the stars and make contact with our interplanetary neighbors. For now, it's thrilling enough to be taking baby steps in that direction from the surface of Mars.

Bonneville crater

sol 50

sol 100

Gusev Crate

Landing Site
sol 1: Jan. 4, 2004

sol 125

Gusev Crater Plains

——— Spirit rover traverse path

0   500 m

0   0.3 miles

North

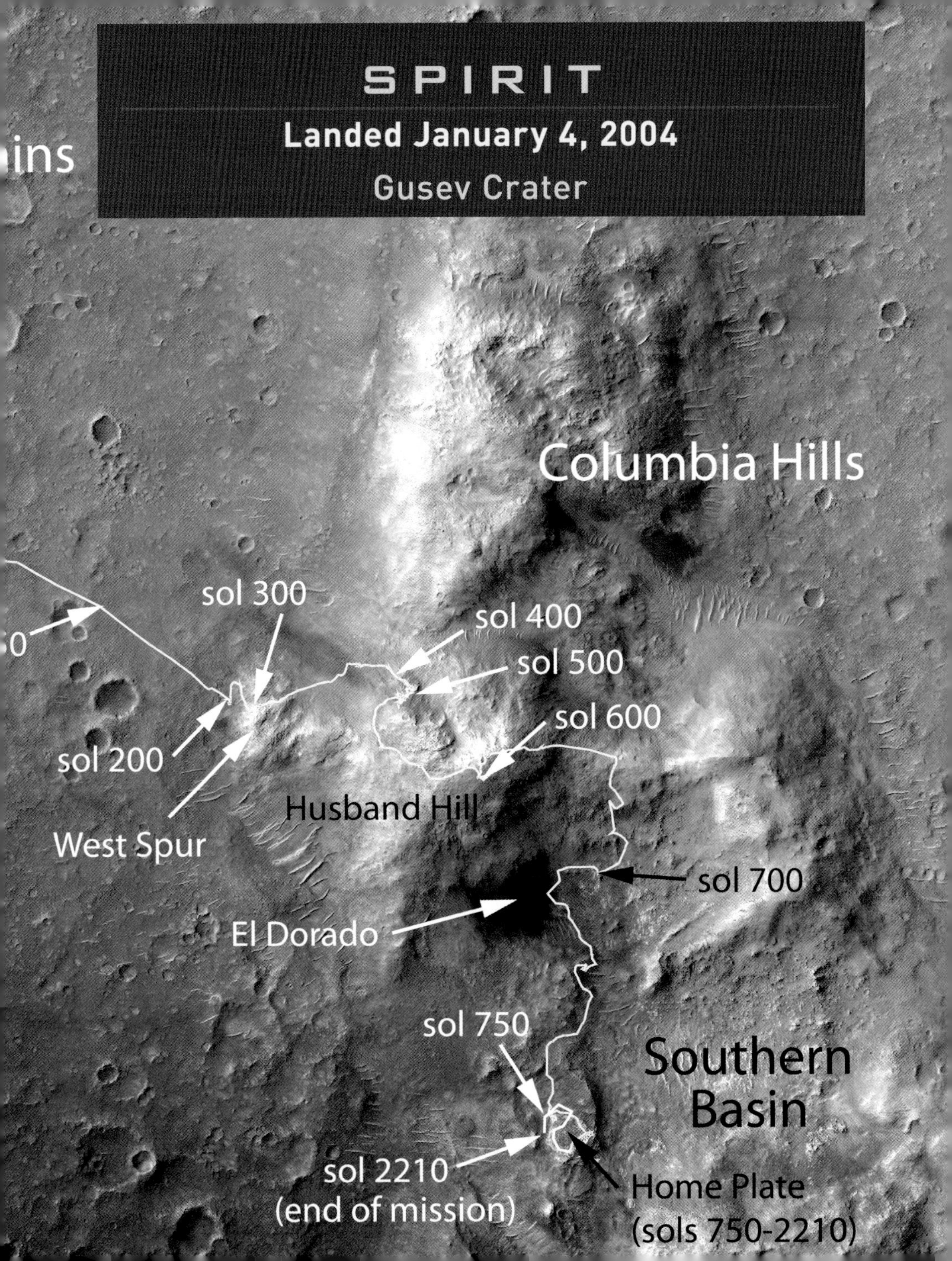

SPIRIT
Landed January 4, 2004
Gusev Crater

ins

Columbia Hills

sol 300

sol 400

sol 500

sol 600

sol 200

West Spur

Husband Hill

sol 700

El Dorado

sol 750

Southern Basin

sol 2210
(end of mission)

Home Plate
(sols 750-2210)

Six Wheels on the Ground

ABOVE Pancam color photo of the *Spirit* lander taken a few sols after egress (sol 16). The deflated airbags that surrounded the lander during the landing on Mars, once pristine and white, are now reddish and dirty after having bounced around so much on the dusty surface. Hundreds of people helped design and build the lander, and seeing it there on Mars was a source of great pride to everyone involved.

OPPOSITE Twelve Martian days after landing (sol 12), *Spirit* rolled down one of the lander's cloth ramps and finally touched down on the surface of Mars. The 3-D image opposite, merged from the left and right rear hazard avoidance camera (Hazcam), shows the results of the rover's successful egress onto the dusty soil. The rover's rear wheels, each about 25 cm (10 inches) tall, can be seen in the lower foreground.

Dust in the Wind

ABOVE *Spirit* sol 12 Pancam color image. The area shown here is a typical piece of Martian ground about 30 cm (1 foot) wide. Volcanic rocks are scattered about the scene. The 3-D dust image opposite was taken in the smooth patch just to the right of the center of this image.

OPPOSITE *Spirit* sol 15 Microscopic Imager 3-D view of a tiny 3 cm (1.2 inch)–wide patch of fluffy, clumpy Martian dust near the lander. Individual dust grains are microns wide (about the size of smoke particles) and thus too small to resolve with the microscope. But clumps and mini-towers of aggregated dust grains can be made out in the image. This kind of porous surface, formed by the gentle settling of dust out of the air, is sometimes referred to as "fairy castle structure."

Adirondack

ABOVE Adirondack was also photographed in color by Pancam (on sol 14) in order to identify the best place to brush and grind it with the RAT. From a study of these kinds of pictures, it became clear that the team would need to use the RAT to grind through the rock's dust coating to determine the intrinsic character of the rock itself.

OPPOSITE This is a *Spirit* rover Front Hazcam 3-D view, taken on sol 15, of the football-sized rock named Adirondack. The rover took several days to position itself such that the rock was right between the front wheels, in the "work volume"—the perfect spot for the rover's arm to reach out and detect the rock's chemistry and other properties. Rover drivers need 3-D views like this to generate a computer model of the terrain and ensure that the rover can place its instruments on a target safely.

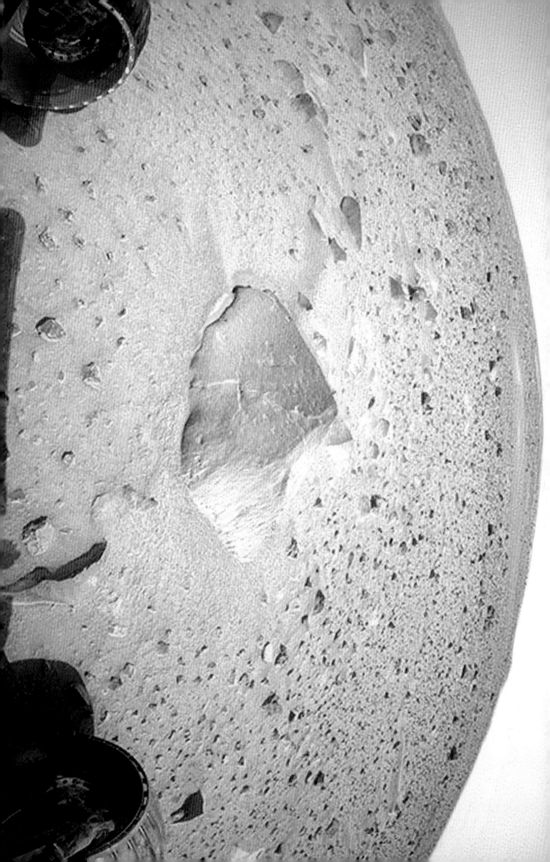

Shores of Bonneville

ABOVE This is part of a large Pancam color panorama of Bonneville Crater acquired on sols 68 and 69. The interior and floor of the crater exhibit dunes and wind streaks but no evidence of the layering or color variations that might indicate different rock types with depth. Thus, despite the spectacular view, exploring Bonneville ended up being somewhat disappointing because *Spirit* never found the ancient buried lake deposits scientists were hoping to discover there.

OPPOSITE *Spirit* took this 3-D Pancam panorama of 200 meter (650 foot)–wide Bonneville Crater when it arrived at the rim on sol 67. The rover had to scramble over large boulders and rugged terrain to reach this dramatic vista point. Barely visible on the far left rim of the crater are the crashed remains of the heat shield that had protected the rover during its fiery entry into the Martian atmosphere.

Dusty Dunes

ABOVE *Spirit* Pancam sol 73 color photo of the "scuff mark" created by the rover wheel in the dune. Dunes like this have been seen throughout Spirit's journey through Gusev Crater, and orbital photos of other places on Mars indicate that dunes are quite common across the planet.

OPPOSITE On sol 73, *Spirit* acquired this 3-D Microscopic Imager photo of a 3 cm (1.2 inch)– wide piece of a small, dusty sand dune on the Gusev Crater plains. The rover had dug into part of the dune with its wheel in order to study the materials inside. The sand grains seen here are from the inside of the dune, and their sizes and shapes indicate that the wind was very active in its formation. This originally black-and-white view has been colorized to simulate the approximate natural color of the scene.

Boulder Ridge

ABOVE This sol 87 *Spirit* Pancam false color image shows the same boulder-covered ridge. Brighter, probably dust-coated rocks are mixed with darker, less dusty rocks in this jumbled terrain. Some of the rocks may have been thrown here as "ejecta" from the impact that created nearby Bonneville Crater, and others may be ejecta from more distant impact events. Still, all of them appear to be bone-dry volcanic rocks.

OPPOSITE This is a *Spirit* Pancam 3-D image of a rock-covered ridge on the Gusev plains drive from Bonneville Crater toward the Columbia Hills (sol 87). Some of the boulders seen here are 1 to 2 meters (3 to 7 feet) across, presenting significant challenges to driving the rover. Fortunately, stereo images provided the critical information needed for navigating a path around such obstacles. This originally black-and-white view has been colorized to simulate the approximate natural color of the scene.

Hollowed Ground

ABOVE Color views of the hollows, like this Pancam mosaic from sol 111, show them to be filled with very fine-grained reddish sand and dust. Apparently, as fine grains move across the surface or settle out of the air from dust storms, they become trapped within small craters and slowly build up over time, creating the relatively flat, bright, dusty hollows.

OPPOSITE *Spirit's* shadow makes for a nice self-portrait in this sol 111 Navcam 3-D image of sand and rocks near a small hollow in the Gusev plains. Stereo imaging of these small circular depressions helped show that they are heavily eroded and partially filled impact craters.

Pot of Gold

ABOVE This *Spirit* Pancam false-color view of Pot of Gold, taken on sol 167, reveals the strange nodular appearance of this potato-sized rock and highlights the less oxidized (bluish in this rendering) qualities of many other nearby rocks and rocky fragments.

OPPOSITE Distant images suggested that the geology might be dramatically different in the Columbia Hills. This hope was vindicated by close-up 3-D images like these from *Spirit's* Microscopic Imager on sols 163 and 170. Here is one of the first radically different kinds of rocks studied by the rover: Pot of Gold (so named because it was "at the end of the rainbow" after the rover's nearly five-month-long drive to the Hills). The surface has small nodules and protrusions, suggesting that substantial physical and possibly chemical weathering has occurred.

Wooly Patch

ABOVE This *Spirit* Pancam color view of Wooly Patch reveals that the color differences between materials inside the rock and those on the surface are not as dramatic as those seen in the Gusev Crater plains (for example, in the RAT hole shown previously from the rock Adirondack). This kind of information supports the idea that these rocks were weathered or altered in some way, probably by water, long ago in Mars' history.

OPPOSITE Another dramatic find in the Columbia Hills was that the rocks there are much softer than the hard volcanic rocks of the plains. This can be seen in these *Spirit* Pancam sol 200 views of twin RAT holes ground into the rock called Wooly Patch. Each hole is about 4.5 cm (1.8 inches) wide and 5.2 mm (0.2 inches) deep. This view has been colorized to simulate the approximate natural color of the scene.

Flanks

ABOVE Pancam color mosaic of
the flanks of Husband Hill, along
the traverse on the West Spur.
Some of the layers and ridges,
seen in the 3-D image, show
slightly different colors in this
view, suggesting that there might
be a change in the chemistry or
mineralogy in those areas. Using
colors to track layered rocks
across large regions is a common
technique used by geologists on
Earth. The rovers have shown that
it can be a useful technique on
Mars as well.

OPPOSITE *Spirit*'s ascent to the top of
the West Spur was only a prelude
to even more climbing, with
the goal of reaching the tallest
peak around—Husband Hill. 3-D
mosaics of the flanks of Husband
Hill, like this one from sols 229
and 230, revealed new evidence
to account for complex layers and
rugged ridges.

First Impressions

ABOVE This sol 263 Pancam color view shows the location of the accompanying 3-D image, for context. The compressed region is within a small "drift" of dust that has gathered between several rocks. Many scientists assume Martian dust is the same everywhere, because the wind mixes it all around. The team periodically tests this assumption by measuring the chemistry and properties of dust piles like this, which are found at different places along the rover's course.

OPPOSITE This *Spirit* sol 240 Microscopic Imager 3-D photo of fluffy, dusty soils shows a donut-shaped indentation mark made by pushing into the dust a 3 cm (1.2 inch)–wide contact plate on one of the rover arm's spectroscopy instruments. Tests like these are sometimes done to try to determine how strong the surface is and how fine (or coarse) the particles are. Here the dust grains are so fine that they compress easily, even under light pressure.

What a Drag

ABOVE This Pancam color view is part of the "Thanksgiving Panorama" taken along the drive to the flanks of Husband Hill from sols 321 to 325, in late November/early December 2004. Spirit's tracks can be followed back more than 100 meters (330 feet) into the distance. What was supposed to have been several days of straight-line driving turned out to look more like a drunken sailor's swagger through these soft sandy soils. Still, despite the slipping and sliding, the rover's software was usually able to compensate and keep *Spirit* on the right path.

OPPOSITE This *Spirit* Navcam 3-D mosaic, looking behind the rover at wheel tracks imprinted in the soft sand, was taken on sol 313 during the rover's traverse from the West Spur to the flanks of Husband Hill. Even though the terrain here is only gently sloped, loose, sandy soil like this still represents a significant potential driving hazard for the rovers.

Craggy Ridge

ABOVE This *Spirit* Pancam sol 392 false color photo shows rocks along the crest of the ridge seen in the accompanying 3-D view. Many of the rocks here are dusty and wind-scoured, but some of them, like the one in the fore-ground, appear much "fresher" and less dusty. This rock and others like it may be relatively younger pieces of debris launched here from distant volcanoes or impact craters.

OPPOSITE Challenging drives along and across steep rocky ridges were a continual theme in *Spirit's* ascent of Husband Hill, which took nearly one Earth year to complete. A dramatic example is seen here in this sol 391 Pancam 3-D mosaic of rocks and craggy ridges approach-ing the summit. Stereo images like this were used to assess the slopes and determine whether the rover and drivers would need to find another way around.

Whale of an Outcrop

ABOVE This is a *Spirit* Pancam sol 500 color mosaic of the Larry's Lookout outcrop. From some angles, the profile of the outcrop resembled the front of a whale, and so this view became known as the "Whale Panorama." Many of these rocks showed signs of having been chemically changed, most likely by water.

OPPOSITE This *Spirit* Front Hazcam 3-D image from sol 490 shows the rover's front wheels nuzzled up to the Larry's Lookout outcrop, near the summit of Husband Hill. The team wanted to reach out with the rover's arm to study the chemistry of the rocks at the top of the outcrop, but the rocks were perched above a layer of soft sand, making the approach difficult. There was also concern that small landslides, like the one in this image, might injure the rover. After several approaches, however, the rover was able to reach the outcrop.

Sentinels

ABOVE This sol 506 *Spirit* Pancam image shows the same rocks in color. The top of the middle rock is very dusty, suggesting that the rocks have been in this static condition for a long time. Details can be seen even in the shadowed parts of these rocks; in general, the Martian sky is so bright that it provides enough light to study the geology of even shadowed regions of a surface.

OPPOSITE This *Spirit* Pancam 3-D image from sol 506 shows numerous details of the shape and texture of three large, isolated rocks encountered on the flanks of Husband Hill, near the summit. The rocks are pitted, perhaps in part from billions of years of sandblasting by the wind, and perhaps in part because they may be volcanic rocks, erupted from a very gassy, bubbly lava flow. The middle rock is about 40 cm (16 inches) across.

Summit

ABOVE This *Spirit* Pancam panorama was taken on sol 582 from the rover's perch at the summit of Husband Hill, looking south toward Tennessee Valley and the Southern Basin of the Columbia Hills. As this and other similar views showed, the summit region is a windy place, and abundant evidence of sand dunes and wind-carved rock faces are visible in high-resolution Pancam images. On several occasions, the winds at the summit cleaned dust off the rover's solar panels (foreground), producing large surges in the rover's solar power supply.

OPPOSITE Finally, after more than 400 sols of climbing, *Spirit* reached the summit of Husband Hill, roughly 100 meters (330 feet) above the plains where the rover landed. This Navcam 3-D mosaic from the summit was taken on sol 582. The rover spent more than 100 sols exploring the rocks, soils, and outcrops along this broad summit plateau.

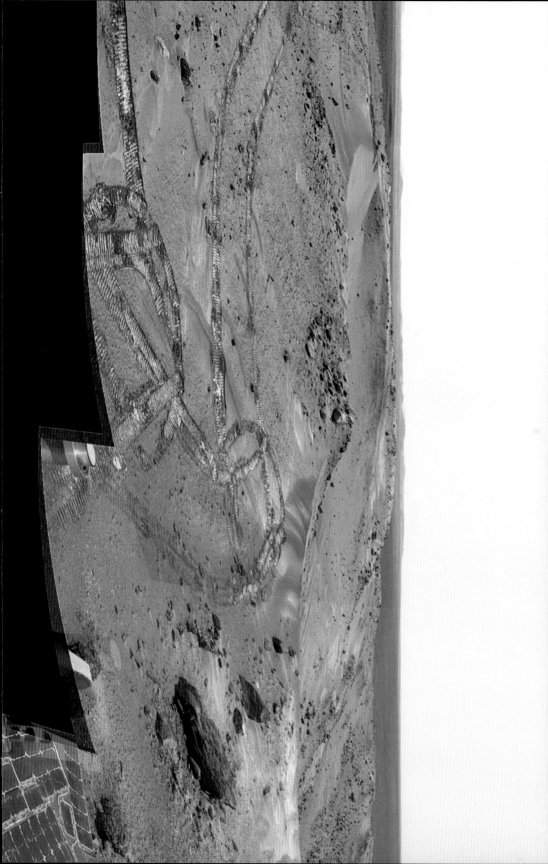

El Dorado

ABOVE This *Spirit* Pancam mosaic from sol 708 shows a broader view of the El Dorado sand dune field. The dunes cover an area about 150 meters (490 feet) across, along the southern flanks of Husband Hill. Rover measurements show that the sand is made up mostly of ground-down pieces of dark volcanic rock (basalt), unlike sand on Earth, which is made mostly of quartz. The relative lack of dust on the sand implies that the dunes may yet be moving, albeit at a very slow rate.

OPPOSITE Mars is covered in sand, and here's what it looks like up close, in this *Spirit* Microscopic Imager sol 709 3-D photo from the "El Dorado" sand dune south of the summit of Husband Hill. The donut-shaped indentation mark seen here was made by pushing a 3 cm (1.2 inch)–wide contact plate on one of the rover arm's spectroscopy instruments into the sand.

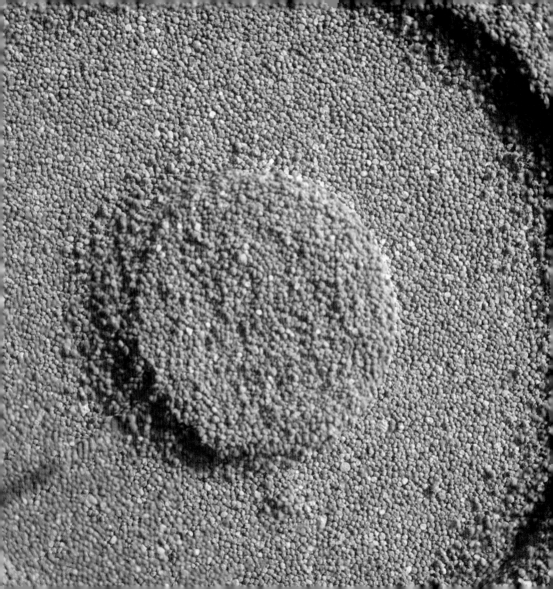

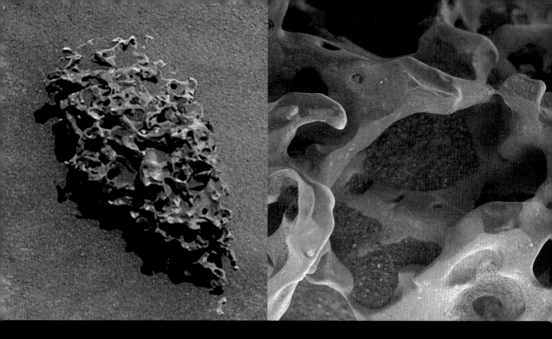

Gong Gong

ABOVE This *Spirit* Pancam sol 736 color photo (above left) shows that what is left of the heavily eroded surface of Gong Gong is mostly dark, clean volcanic rock (basalt). However, the numerous small pits and mini "caverns" in the rock have trapped dust in some places. The spectacular microstructure of the rock is beautifully seen in the Microscopic Imager view (above right), which shows an area only 3 cm (1.2 inches) across.

OPPOSITE This *Spirit* Pancam sol 736 3-D photo shows one of the most dramatic examples of rock texture ever seen on Mars. This small piece of rock, about 17 cm (7 inches) long and called "Gong Gong," is a heavily eroded piece of what geologists call scoria: a pitted, vesicle-covered, wind-carved volcanic rock. There must have been a high concentration of gas in this lava, judging by all the frothy bubbles that were formed when it erupted.

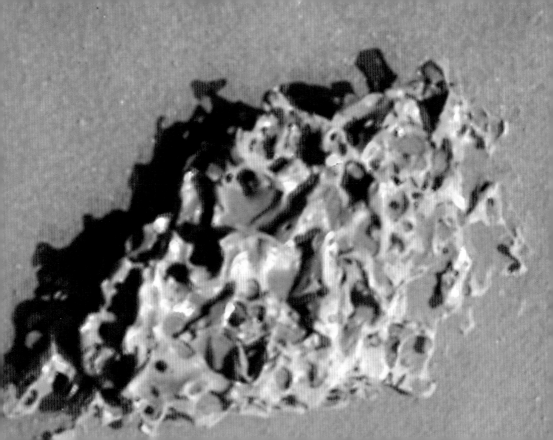

Home Plate

ABOVE This *Spirit* Pancam sol 751 false color view shows a close-up of some of the fine layers along the edge of Home Plate. Individual layers are a few centimeters (about one inch) thick here. One of the layers appears to sag under the weight of a small embedded rock. This kind of feature is often called a "bomb sag" by geologists, and may result from a hot volcanic fragment erupting onto soft, perhaps even wet, volcanic ash deposits.

OPPOSITE After many months of descending the southern flanks of Husband Hill and working its way into the Southern Basin, *Spirit* finally arrived at the bright, circular feature named Home Plate. This sol 748 Pancam 3-D mosaic shows the edge of Home Plate, which turned out to be made of finely layered stacks of rocks, about 1.5 meters (5 feet) thick. The layers could be the result of sedimentary, volcanic, or even impact cratering processes.

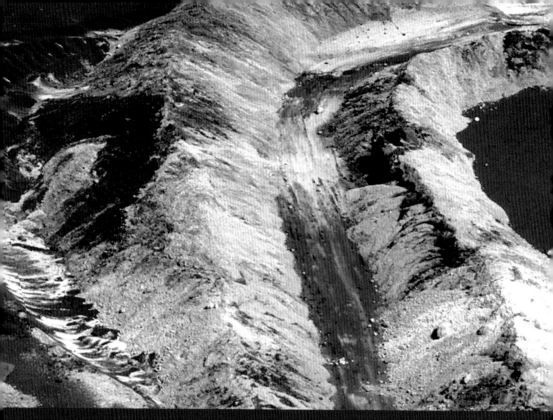

Salty Soil

ABOVE This Pancam sol 788 color view shows that there were several shades of bright soil dug up by the stuck wheel. Chemical analysis of the bright soils shows them to be made of concentrated, sulfur-rich salty minerals. It's still a mystery why these salty deposits are buried just below the surface near Home Plate. Sulfur-rich volcanic gases may be slowly percolating through and weathering the rocks, or they may be the evaporated remnants of salty liquid water.

OPPOSITE The most troublesome sign of "old age" on the *Spirit* rover by the time it reached Home Plate was the failure of the right front wheel motor. The team had to drive the rover backward, dragging that stuck wheel through the dirt. Luckily, dragging the wheel created a 15 cm (6 inch)– wide trench and dug up some bright, buried soils that we never would have seen otherwise, depicted in this sol 788 Pancam stereo view.

Madeline English

ABOVE This sol 1163 Pancam false-color view shows details of the "Madeline English" outcrop along the easternmost edge of Home Plate. The layering and frothy looking rock chips nearby suggest a volcanic origin. For scale, the rocky slab at upper left is about 5 centimeters (2 inches) thick. Dark sand deposits, forming lovely ripples and wind tails, have been eroding this outcrop for billions of years.

OPPOSITE Details in the finely-layered rocks exposed at Home Plate provide geologists with clues about the origin of this interesting geologic structure, which *Spirit* studied for most of its mission. Here, layering in the brighter outcrop material can be seen down to the limit of resolution of this Pancam 3-D view, consistent with, for example, multiple episodes of volcanic ash eruptions forming these rocks.

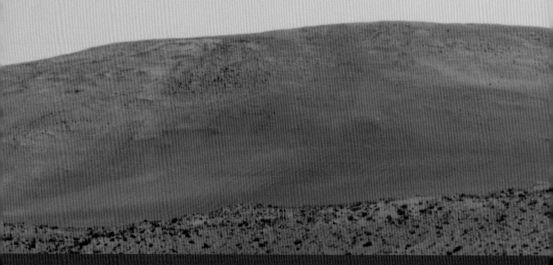

West Valley

ABOVE This high resolution view of Husband Hill and the El Dorado sand dune field was taken from atop Home Plate. Every so often the rover would be programmed to take photos of El Dorado to see whether the sand was changing over time. Here, the left half of the dune field appears brighter and redder than at other times, suggesting that it got dusty. Such changes at El Dorado were monitored for more than three Earth years.

OPPOSITE While driving across the Western edge of Home Plate plateau on sol 1369, *Spirit* obtained its best view of the valley to the west. Home Plate is about 2 meters (6.5 feet) above the surrounding plains. A small, likely volcanic hill named Tsiolkovski can be seen in the left mid-field, and Husband Hill and the El Dorado sand dune field can be seen in the distance at right.

Roadcut

ABOVE The corresponding Pancam false color view of layered rocks along the edge of Home Plate shows that despite what has likely been 3 to 4 billion years of erosion, its basic structure—a layered, circular plateau—remains intact. This suggests to geologists that the layers are strong because they were compressed and cemented together over time, perhaps by hydrothermal weathering from ashy volcanic cinder cone eruptions.

OPPOSITE As *Spirit* descended from the Home Plate plateau and continued to circle the feature on its western side, the views of the edge of the structure became more dramatic. In the sol 1886 3-D mosaic, taken at a place called Troy, not far from *Spirit's* final position, layered rocks from what may have been the lower parts of an ancient, eroded volcanic cinder cone rise up more than 2.5 meters (8.2 feet) above the rover.

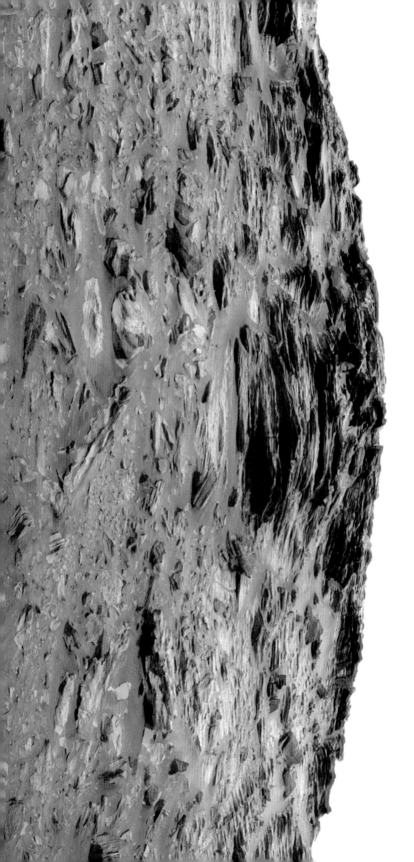

Von Braun Hill

ABOVE This sol 2114 false-color Pancam view of Von Braun hill is from one of the last few sets of color panoramas taken by *Spirit*. Enigmatic and capped with resistant, perhaps Home Plate-like material, Von Braun would have been *Spirit's* next drive target. However, communication was lost with the rover on sol 2210 because of low winter solar-power levels and several wheels getting stuck within Solander crater.

OPPOSITE As the *Spirit* rover continued around the western border of Home Plate near the end of 2009, two new features came into view: a light-toned, round patch of soil known as Solander crater, and a more distant cone-shaped hill known as Von Braun. Unknown to the rover team until it was too late, Solander crater was found to be filled with soft, sandy material that would ultimately trap Spirit and end the rover's mission.

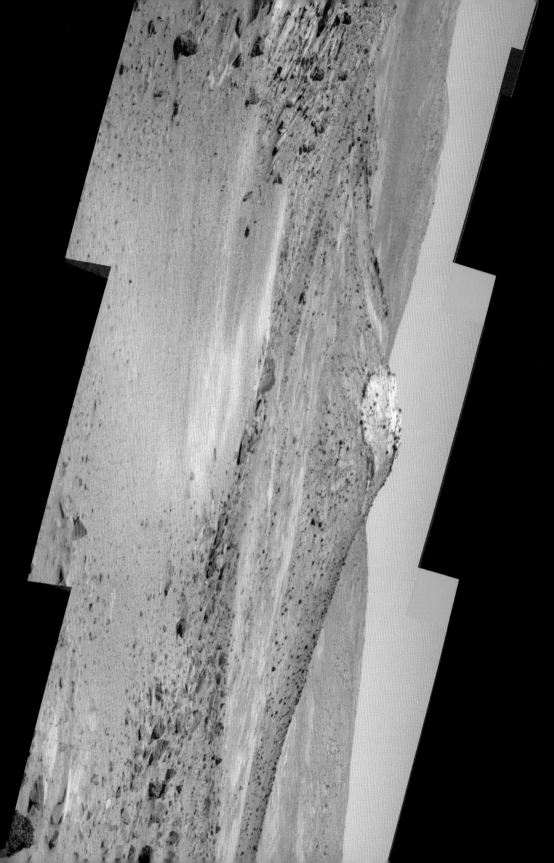

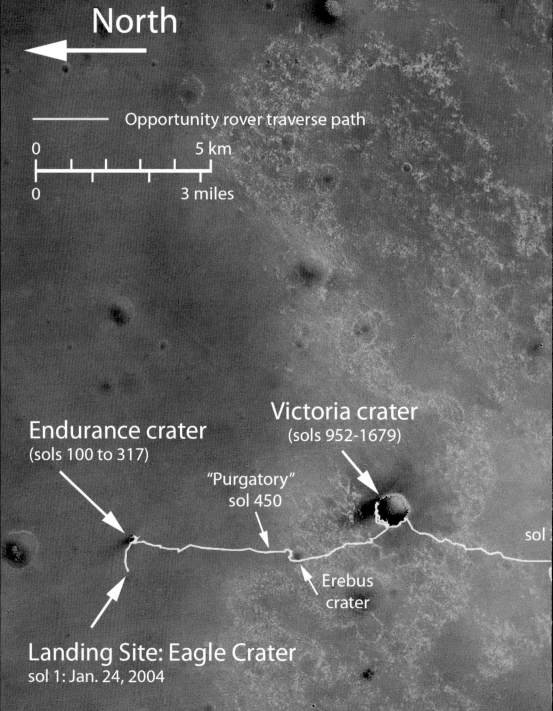

North

Opportunity rover traverse path

0   5 km

0   3 miles

Victoria crater
(sols 952-1679)

Endurance crater
(sols 100 to 317)

"Purgatory"
sol 450

sol

Erebus
crater

Landing Site: Eagle Crater
sol 1: Jan. 24, 2004

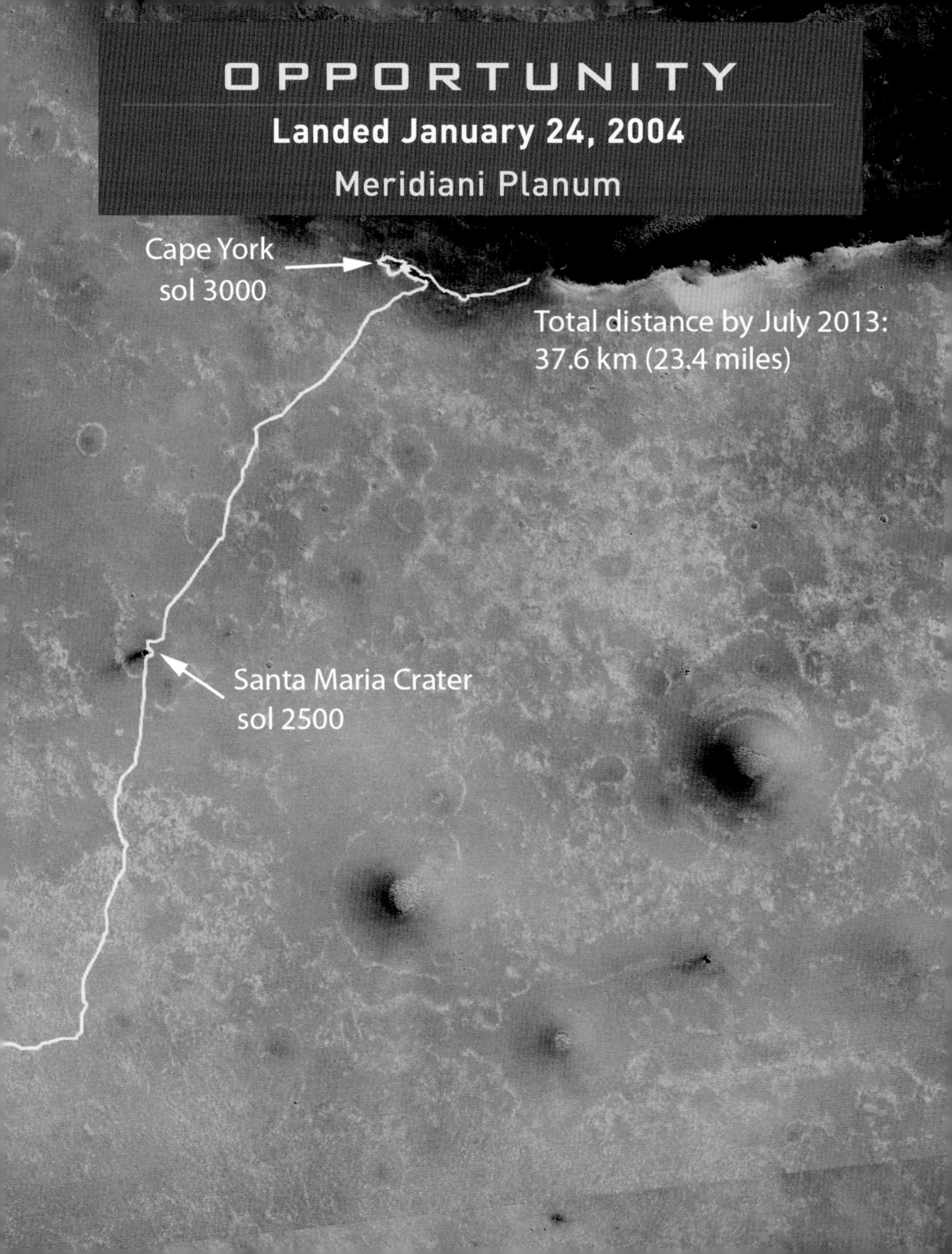

OPPORTUNITY
Landed January 24, 2004
Meridiani Planum

Cape York
sol 3000

Total distance by July 2013:
37.6 km (23.4 miles)

Santa Maria Crater
sol 2500

Eagle Crater

ABOVE This Pancam color mosaic of Eagle Crater was also shot on sols 58 to 60. The Eagle Crater landing site is dramatically differ-ent from the Gusev Crater landing site. For example, the surface has a dark, more brownish color than Gusev's redder, dustier soils. The small amounts of dust visible here occur mostly in small wind-blown drifts, or in a bright dusty "tail" around the east (right side) edge of the crater.

OPPOSITE The *Opportunity* lander bounced across the flat plains of Meridiani Planum and came to rest here in the middle of Eagle Crater, a small, shallow 20 meter (66 foot)–wide, 1.5 meter (4.9 foot)–deep impact crater. The rover spent nearly two months driving inside the crater and studying the fascinating outcrop rocks seen poking out of the crater's far inside wall. This *Opportunity* Pancam 3-D mosaic was shot on sols 58 to 60.

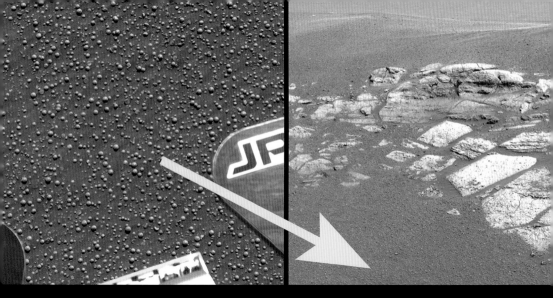

Blueberry Fields

ABOVE In color images of these grains, team members noticed that they are grayer and less red than the dark, reddish-brown soils they rest on. This color difference is dramatically enhanced in false-color images, where the grains, dubbed "blueberries," burst out into bluish hues. This high-resolution false-color Pancam image (above left) shows a typical blueberry-filled scene on the floor of Eagle Crater. Above right is a wider-angle false-color image looking out of the crater and across the plains.

OPPOSITE One of the most stunning and surprising discoveries made by the *Opportunity* rover is that the surface of Meridiani Planum is covered with countless millions of small, spherical rock grains, like those seen here in this Microscopic Imager 3-D view from sol 19. The field of view here is about 3 cm (1.2 inches), and the spherical grains range from about 2 to 4 mm (0.08 to 0.16 inches) in diameter.

Layers and Berries

ABOVE Here, a special combination of Pancam's ultraviolet, visible, and infrared filters was used to construct a false-color "berry finder" map. Blueberries scattered across the ground, as well as those sitting on and embedded in the outcrop rock, light up as hot pink. Later imaging, chemical, and mineral measurements taken of the blueberries showed them to be rich in iron, probably formed in the same way that some minerals grow during the slow evaporation of liquid water.

OPPOSITE There was obviously some connection between the blueberries and the outcrop. This sol 15 stereo image of a region of blueberry-filled layered outcrop rock in Eagle Crater helped solve the mystery. The berries occur naturally inside the layers of the bright, reddish outcrop rocks. The bright rock is easy for the wind or other forces to break apart and erode away, allowing the berries to pop out of the rock and collect on the surface.

Eagle Trench

ABOVE Again using the Pancam's false-color filters, trenches like this can be investigated to see if a layer of blueberries extends underground. Looking at the trench walls in this sol 26 mosaic, it's hard to see any evidence of blueberries more than a few centimeters (an inch or so) under the surface. Such pictures helped to show that the blueberries make up just a thin layer on the uppermost surface of the soils in Meridiani Planum.

OPPOSITE The team needed to discover what lay below the blueberries and soil. The rovers don't have digging or drilling tools, but they can make shallow trenches in the soil using their wheels. This trench, excavated into the sandy soils of Eagle Crater on sol 26, is about 15 cm (6 inches) wide and 10 cm (4 inches) deep. The degree of layering of the subsurface soils and the sharpness of the trench walls provide unique information about the physical properties of the Meridiani soils.

Ratted Berries

ABOVE This sol 36 *Opportunity* Pancam false color image of the McKittrick RAT hole shows that the fine powdery rim outside the hole, which is made up partly of ground-up blueberries, is brighter and redder than the unground outcrop rock. This was a clue that the berries contain iron minerals, which often get brighter and redder when they are ground up into smaller sizes.

OPPOSITE *Opportunity's* Rock Abrasion Tool (RAT) has been essential for understanding the history of this enigmatic landing site. On sol 30, the RAT was used to grind a 5 mm (0.2 inch)–deep hole into the Eagle Crater outcrop rock "McKittrick." The RAT ground down into two blueberries (lower middle, lower right), exposing their interiors in cross-section. None of the berries "shaved" this way showed evidence of internal layering or external coatings, further supporting the idea that they had been slowly "grown" from the evaporation of ancient, salty water.

Dangerous Driving

ABOVE The beautifully layered rocks seen on Endurance's inner rim wall near the upper center of this view, named "Burns Cliff," were quickly identified as the most interesting and accessible geologic target for study. It would be impossible to navigate the rover over the steep (60 to 90 degree) slopes, like those seen on the left side of this view. Instead, the drivers focused on the more gentle (10 to 25 degree) slopes, seen on the distant rim on the right side of this view.

OPPOSITE Endurance Crater, at 130 meters (about 426 feet) wide and 20 meters (66 feet) deep, had steep slopes that made for spectacular vistas but potentially dangerous driving conditions. One wrong turn or bad slip across a slick berry-covered rock could mean disaster if the rover went over the edge. This Navcam 3-D image was taken near the edge of the rim of Endurance Crater on sol 115.

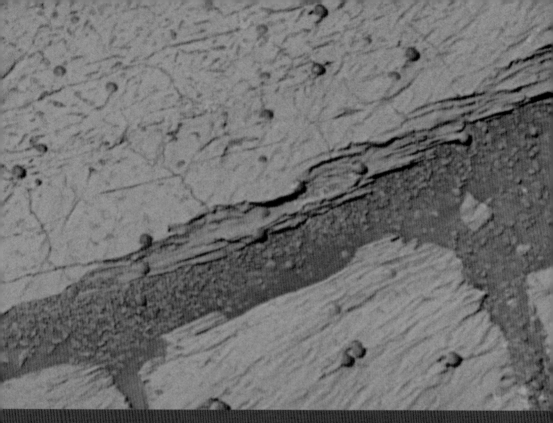

Embedded Berries

ABOVE This broader sol 140 Pancam
color view of blueberries poking
out of outcrop rocks within
Endurance Crater shows small
"wind tails" growing downwind of
the berries, as the outcrop rock
surrounding them slowly erodes
away. This is one way that scientists can deduce the modern-day
average wind direction in this
region. These kinds of surfaces
are also potential driving hazards,
because the rover's wheels could
slip downhill if some of the berries
get loosened.

OPPOSITE During the descent into
Endurance Crater, the team
used the microscope extensively
to study the textures of different layers inside the crater. This
Opportunity Microscopic Imager
stereo image of a 3 mm (0.1
inch)– wide iron-rich blueberry
embedded within sedimentary
outcrop layers was shot from
about 2 meters (6.5 feet) inside
Endurance Crater on sol 142. The
layers are extremely fine—so
much so that a geologist would
refer to these as "laminations"
rather than layers.

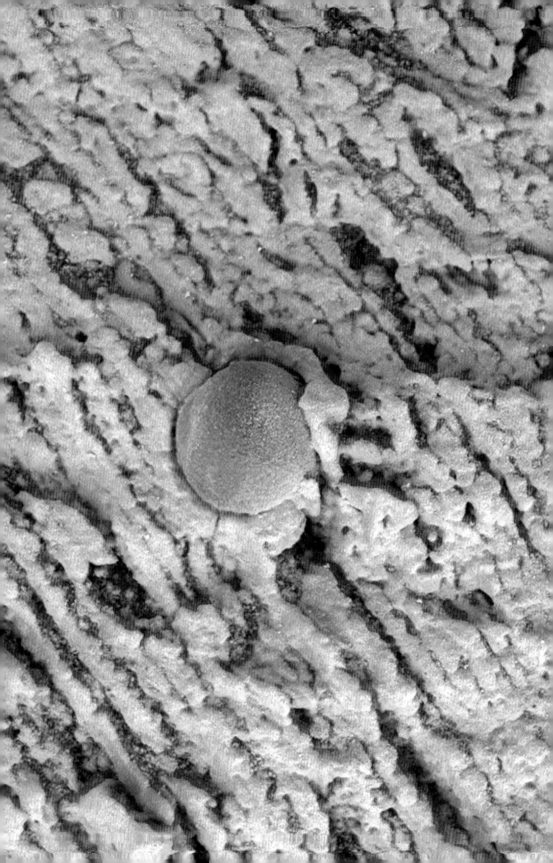

Swiss Cheese

ABOVE During parts of the descent into Endurance Crater, the texture of the layers over which the rover drove changed dramatically across very short distances. This Pancam false color image from sol 150 shows three RAT holes that were ground into areas only 5 to 20 cm (2 to 8 inches) away from each other. Characterizing the chemistry, mineralogy, and physical properties of the layers provided the information needed to try to reconstruct the complex geologic history of the crater, along with the pre-crater surface.

OPPOSITE This is a sol 152 *Opportunity* Microscopic Imager 3-D image. It shows part of a 4.5 cm (1.8 inch)–diameter RAT hole ground into a very soft sedimentary rock layer about 3 meters (10 feet) along the traverse down into Endurance Crater. This rocky layer is very porous. Several holes or sockets can be seen (most prominently at center right) where blueberries popped out during the RAT grinding.

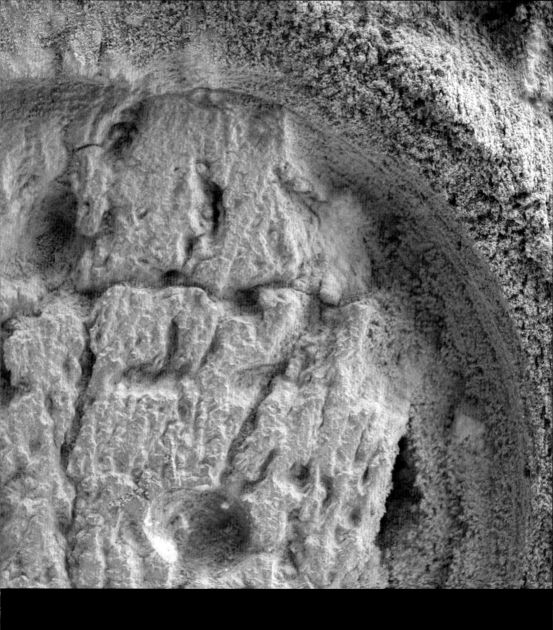

Flaky

ABOVE Pancam color data from sol 170 didn't reveal any dramatic differences in color between the flaky veins and the surrounding host rock, suggesting that the flakes and the host rock may be of essentially the same material. Rover arm chemical measurements on some of the flakes suggest that water-formed mineral rinds may be the best explanation for these features.

OPPOSITE At several points inside Endurance Crater, *Opportunity* encountered what appeared to be thin, flaky mineral veins growing within rock fractures. The small fractures seen here in this Pancam stereo view are only about 20 cm (8 inches) long. These veins may have been the result of groundwater forming mineral-rich rinds once it evaporated. Or they could represent sand or other mineral grains that flowed down and filled the cracks from above and that were somehow cemented together into the more resistant veins.

Upslope

_{ABOVE} This false-color *Opportunity* Pancam mosaic from sol 175 is an example of a typical "side survey" imaging sequence used to try to characterize the textures and colors of layers encountered by the rover during the descent into Endurance Crater. Several RAT holes ground into some of the different layers can be seen in the upper right of the image.

_{OPPOSITE} *Opportunity's* Navcam took this dramatic stereo mosaic on sol 171 while climbing down Endurance Crater and tilted downhill nearly 30 degrees. The view near the center of the image is toward Burns Cliff, a 2.5 meter (8 foot)–high stack of spectacularly layered rocks, about 80 meters (260 feet) away from the rover. To the right, the view is that seen toward the rear of the rover, looking up toward the steep slope down which the vehicle had been descending.

Berry-Free RAT Hole

ABOVE This sol 180 Pancam color
view of the Diamond Jenness
RAT hole shows how carefully the
rover drivers had to work. They
had to position the rover and the
arm instruments on the "clean-
est" part of the outcrop layer to
avoid letting any of the blueberries that had fallen from layers
above contaminate the measure-
ment. The RAT hole is about 4.5cm
(1.8 inches) in diameter.

OPPOSITE As the rover continued into
Endurance Crater, the outcrop
textures began to change. Close to 20
meters (66 feet) along the drive down,
Opportunity used its RAT to grind
into what turned out to be a very soft
layer of outcrop rock called "Diamond
Jenness." When the Microscopic
Imager took this 3-D image of the
resulting 7 mm (0.3 inch)–deep RAT
hole, there were some interesting
discoveries: no blueberries! Perhaps
this layer was formed at a time
before the berries had been formed
in other rocks, or perhaps the berries
had eroded away.

Field of Dunes

ABOVE On Mars, giant craters like Endurance are also big sand traps. Over time, if a supply of sand remains available, more and more sand is trapped and these central crater dunes grow to essentially fill the entire bowl of the crater. Other craters encountered by *Opportunity* have been completely flattened out this way.

OPPOSITE The enormous dune field at the bottom of Endurance Crater was a tantalizing but dangerous target for *Opportunity*. We were as close as we'd ever been to a large collection of dark sand dunes—and we wanted to measure them in detail. However, as this sol 201 Pancam stereo image shows, the sand is a potential rover death trap. Some dunes were more than a meter (over 3 feet) tall. Opportunity would almost certainly get stuck. Fortunately, the rover was close to several dune tendrils, and so could study the material without actually going in.

Wopmay

ABOVE This Pancam sol 251 color view of Wopmay shows some of the best examples of the lumpy textures and polygonal cracking seen on the surface of this rock. Wopmay could have fallen down from a layer higher up in the crater, or could be a block of ejecta that just happened to land here after the impact. Evidence that the rock originated from deeper layers comes from closer-up photos that show a relative lack of blueberries on its surface.

OPPOSITE Nearing the lowest point of the descent into Endurance Crater, *Opportunity* encountered a number of strange rocks with interesting and unique textures. One such example is the 1 meter (3.3 foot)–wide, 2 meter (6.6 foot)–long rock named "Wopmay," seen here in this Navcam sol 250 stereo image. Wopmay has a lumpy surface covered with polygon-shaped cracks, suggesting that its surface has perhaps been heavily weathered, or may even have undergone cycles of wetting and drying.

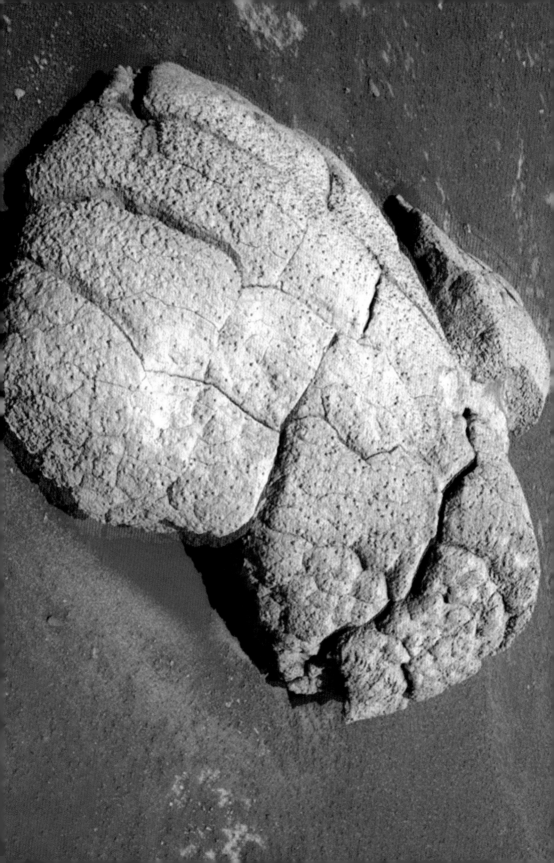

Racetrack

ABOVE This *Opportunity* Pancam sol 287 color image shows additional details in the layering as the rover approached Burns Cliff. These layers are what geologists call "crossbeds," and they are most likely the lithified, exposed interiors of what used to be sand dunes on the Martian surface. The way some of the crossbed layers intersect one another is typical of exposed ancient sand dune cross-beds found on Earth.

OPPOSITE *Opportunity's* ultimate destination within Endurance Crater was Burns Cliff, a stack of spectacularly layered sedimentary rocks. Approaching the cliff was tricky. The rover drivers had to use stereo images and results from the test rover in California to plan *Opportunity's* drive across a 20 to 30 degree slope. This Pancam sol 287 stereo view approaching Burns Cliff makes the terrain seem relatively flat, until you realize that the rover and the cameras are themselves tilted at more than a 20-degree angle.

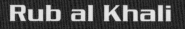

Rub al Khali

ABOVE This *Opportunity* Pancam photo is part of the "Rub al Khali" 360-degree panorama acquired on sols 456 to 464 from the plains south of Endurance Crater. Rub al Khali means "empty quarter" in Arabic, and the name is certainly fitting for this stark and lonely landscape. This panorama was taken while the rover was stuck here, after plowing itself into a slightly taller sand ripple. It took six weeks of spinning the wheels backward to extract the rover.

OPPOSITE This is what the typical scenery in Meridiani Planum looks like. Meridiani is basically flat; the mini-dunes seen in this sol 125 Navcam stereo mosaic are only about 5 to 10 cm (2 to 4 inches) high from crest to trough. The heights, specific shapes, and spacings of these ripples all provide information on the strength, direction, and duration of the winds in this part of Mars.

Purgatory

ABOVE After spinning its wheels backward for six weeks, *Opportunity* finally got free of what had come to be known as "Purgatory Dune" because the team wasn't sure whether the rover would ever escape. This colorized Navcam sol 491 photo shows the old (faded) wheel tracks going into the ruts, and the newer (sharper) tracks coming out.

OPPOSITE This *Opportunity* Pancam sol 496 stereo image shows the 15 cm (6 inch)–wide, 25 cm (10 inch)–deep rut that was created when the rover got stuck in the soft sands north of Erebus Crater. The rover "plowed into" this little dune at the breakneck speed of about 4 cm per second (0.1 mph). To the rover, the dune wasn't really an obstacle: The wheels kept happily spinning, and there was no indication that anything was wrong.

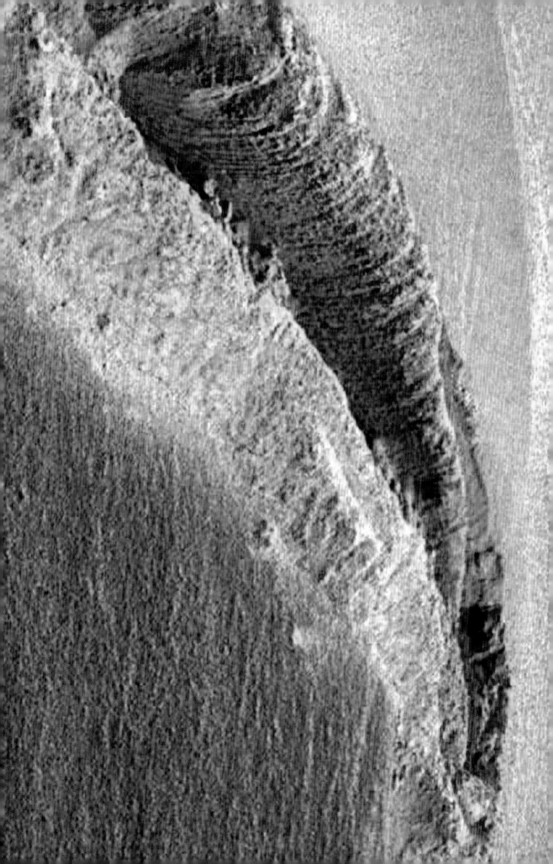

Heat Shield

ABOVE The remains of the heat shield shining in the sunlight generated camera glints off the metallic foil surfaces in this sol 335 Pancam color view. The crash site is littered with mechanical debris like springs, bolts, and metallic shards.

OPPOSITE During the landing in January 2004, *Opportunity* was protected from the fiery heat of atmospheric entry by a heat shield. Just before landing, the heat shield was jettisoned and crashed onto the surface far from the landing site. This *Opportunity* sol 335 Pancam stereo image shows the remains of the heat shield in high-resolution detail. The shield was originally conical but the impact appears to have flipped it inside out; the metallic foil layer of thermal blanketing seen here used to be the inner surface.

An Ocean of Sand

ABOVE In many places, the sands of Meridiani Planum are organized into exquisite patterns. This Pancam southward-looking sol 447 color mosaic, for example, reveals interesting fine-scale details within low sandy ripples near Purgatory Dune. The wavy layering parallel to the ripple ridge near the center of the photo indicates that the ripples themselves are probably internally layered or sorted into different particle sizes.

OPPOSITE Sometimes driving *Opportunity* across the sandy plains of Meridiani Planum felt more like sailing a ship across a vast wavy sea. This Pancam sol 433 stereo mosaic looks south from the rover's position near Voyager Crater, about 1.8 kilometers (1.1 miles) south of Endurance Crater. The terrain is so flat here that the low sandy ripples are the only topographic features to be seen, all the way out to the horizon, which is more than 3 kilometers (approximately 2 miles) away.

Strawberry RAT Hole

ABOVE This *Opportunity* Pancam sol 561 false-color image shows the accompanying RAT hole, as well as a second RAT brush spot, on the rock "Strawberry," located near Erebus Crater. The rover arm had to be carefully positioned in order to sample parts of the rind as well as the "normal" outcrop, so that differences in their respective chemistries could be determined.

OPPOSITE Crossing the boundary from the darker plains north of Erebus Crater to the lighter plains to the south also saw an accompanying change in the texture and chemistry of the sedimentary outcrop rocks. Some of the rocks started showing evidence of thick rinds or coatings. For example, this *Opportunity* Microscopic Imager sol 558 stereo mosaic (inset) shows a 4.5 cm (1.8 inch)–diameter, 5 mm (0.2 inch)–deep RAT hole ground into a thick coating that occurs on top of soft outcrop rocks near the rim of Erebus Crater.

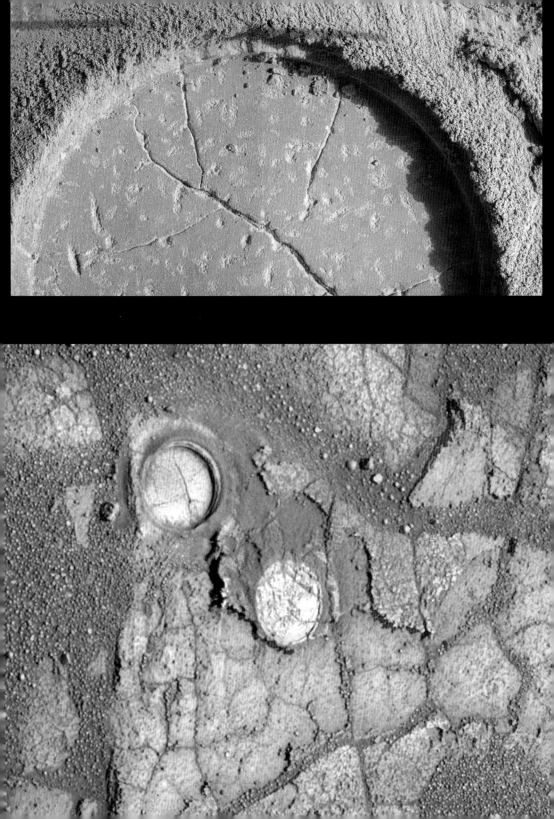

Stacks

ABOVE Images using the Pancam's ultraviolet and infrared filters can provide unique information on subtle physical or compositional differences among outcrop rocks. Such images have to be generated as "false-color" views because they represent colors that are not visible to the human eye. Here a natural color (RGB) composite to the accompanying 3-D image (above left) shows that to the human eye, most of the rocks in this region would have roughly the same colors. However, using the ultraviolet and infrared filters to generate a false-color image (above right) reveals different shades of purple, blue, and red in this *Opportunity* Pancam sol 674 view.

OPPOSITE This view shows even more dramatic evidence of fine-scale layering of outcrop rocks near the rim of Erebus Crater. The team was able to take so many detailed images near the rim of Erebus Crater partly because *Opportunity* was experiencing problems with the shoulder joint motor on its arm; as a result, the engineers and rover drivers had to spend many days diagnosing and developing a work-around for the problem.

Accidental Trench

ABOVE An enhanced color composite Pancam photo taken on sol 836 of the accidental trench reveals that the wheels were able to dig underneath the thin surface layer of blueberries to reveal brighter, essentially berry-free soils below. The lag deposit of berries that was first discovered back in Eagle Crater appears to extend over most of *Opportunity's* traverse north of Victoria Crater.

OPPOSITE The experience of getting stuck at Purgatory Dune resulted in the need to come up with new driving methods that would help the team avoid getting the rovers stuck. This turned out to be prudent planning, because around sol 836, the rover got stuck again in a soft, deeper-than-expected sand dune just north of Victoria Crater. This sol 836 Pancam stereo image shows the "accidental trench" dug by the wheels while the rover was slipping in the sand. Fortunately, it took only a few sols to get the rover out of this particular jam.

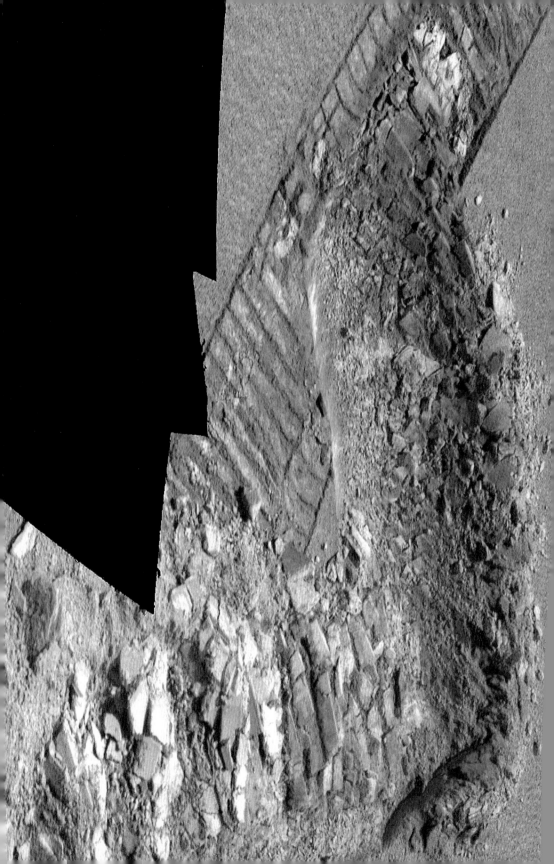

Cape St. Vincent

ABOVE Color imaging has been an important part of the reconnaissance of Victoria Crater. Color views like this Pancam sol 1121 mosaic provide a way to estimate the composition and physical properties of the rock layers remotely.

OPPOSITE After more than two Earth years of driving south, getting stuck in the sand, and having a forced stop to repair a balky arm, *Opportunity* finally made it to Victoria Crater around sol 952. Victoria is the largest crater yet visited by either rover, at 800 meters (0.5 miles) across and 50 to 70 meters (165 to 230 feet) deep. Steep cliffs, 6 to 15 meters (20 to 50 feet) tall flank the crater's rim, exposing a variety of distinct rock layers. This sol 1121 Pancam stereo view of the 6 meter (20 foot)–high Cape St. Vincent promontory is typical of the spectacular new landscapes now being explored by the rover.

Resolution Crater

ABOVE A higher-resolution Pancam color view of the small 6-meter (20-foot)-wide crater Resolution, named after Captain James Cook's eighteenth-century ship of exploration. The impact of a tiny high-velocity asteroid excavated pieces of previously buried subsurface materials. The large number of fresh-looking ejected blocks suggest that the impact event occurred relatively recently in Mars history.

OPPOSITE Along the long traverse from Victoria crater to the rim of Endeavour crater, the *Opportunity* rover team continued to use impact craters as free natural "roadcuts" to explore the chemistry and geology of the subsurface. This sol 1826 3-D view from the Navigation cameras shows the small, fresh crater "Resolution" (upper left), which punctured into surrounding light-toned bedrock.

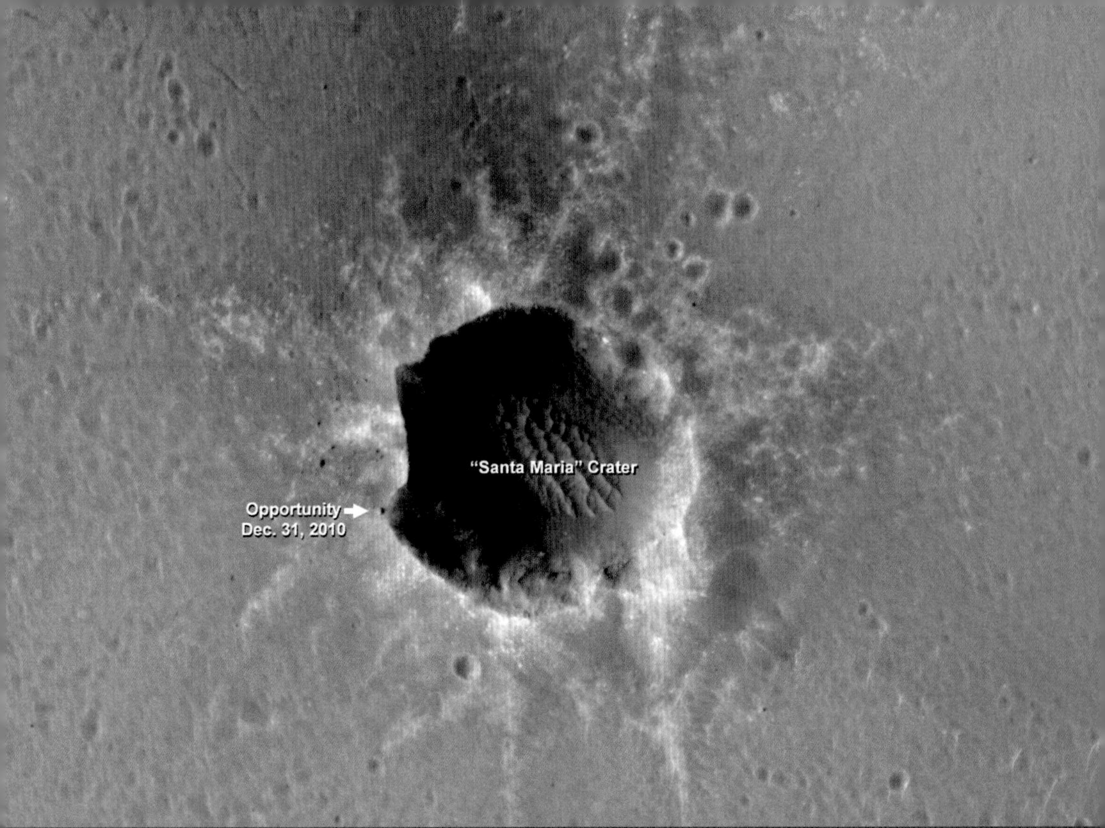

"Santa Maria" Crater

Opportunity
Dec. 31, 2010

Santa Maria

ABOVE Orbital view of Santa Maria crater from the NASA Mars Reconaissance Orbiter HiRISE camera, including a small rocky-looking object on the rim that is actually the *Opportunity* rover. The rover spent three months driving around and studying this impact relatively young and fresh-looking crater, including during a solar conjunction period when Mars was out of communication with Earth for several weeks.

OPPOSITE Another stop on *Opportunity's* long drive to Endeavour crater was at the 90-meter (300-foot)-diameter crater "Santa Maria," seen here in a Pancam stereo view. Some excavated rocks near the rim of the crater appear similar to fine-grained mudstones on the Earth, and may also contain hydrated minerals. Steep slopes and a sandy floor prevented exploration inside the crater, however.

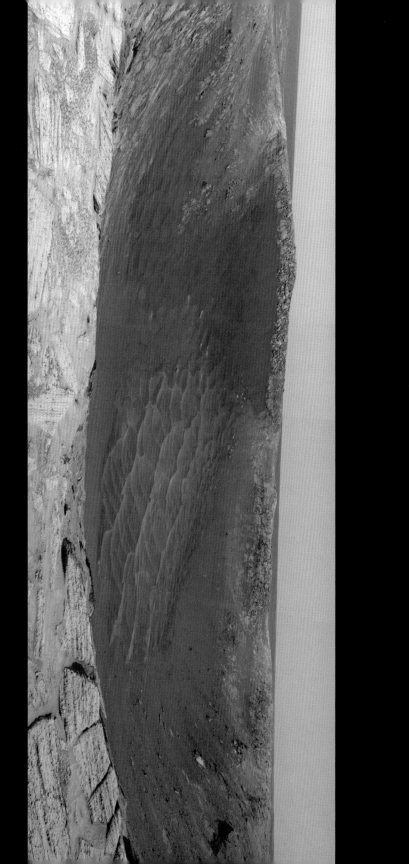

Friendship 7

ABOVE Pancam false-color sol 2586 mosaic of Friendship 7 crater. Named after astronaut John Glenn's 1962 Project Mercury orbital capsule, this small crater shows relatively fresh-looking layered outcrop rocks, similar to those seen previously in Meridiani Planum, ejected from the subsurface. This young crater's interior has been partially filled in with dark sand blowing across the plains.

OPPOSITE Shortly after leaving Santa Maria crater, the *Opportunity* rover encountered the small 16-meter (52-foot)-diameter crater "Friendship 7." This crater is an example of the many kinds of waypoints established along the rover's long traverse in the sandy plains of Meridiani Planum. Craters, fractures, and outcrop exposures help to characterize the geology and chemistry of this part of the planet.

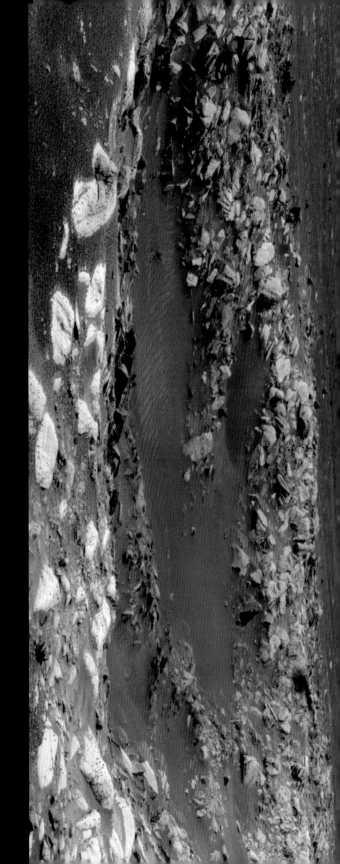

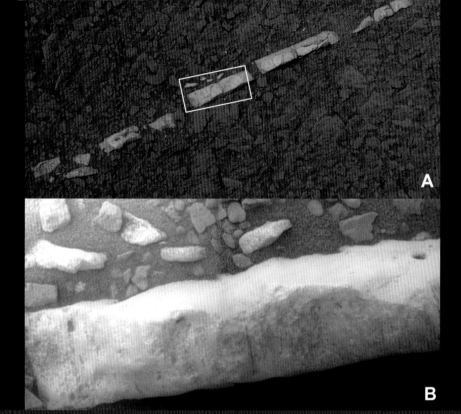

A

B

Jake Matijevic

ABOVE Among the new kinds of features discovered by *Opportunity* at Cape York are 2.5-centimeter (1-inch)-thick bright veins like this one, imaged by Pancam (top) and zoomed in by the Microscopic Imager (bottom). Evidence of calcium, sulfur, and water in these veins suggests that they are made of gypsum, a mineral formed from the interaction of rock and subsurface water long ago, perhaps during the Endeavour impact event.

OPPOSITE The *Opportunity* rover's most recent excursion has been the exploration of the Cape York part of the ancient rim of the enormous 20-kilometer (12-mile)-wide Endeavour crater, where orbital measurements show evidence of hydrated minerals—possibly clays. This sol 3137 3-D view of impact-jumbled rocks and outcrop shows Matijevic Hill, named after our late JPL rover team systems engineer and colleague Jake Matijevic.

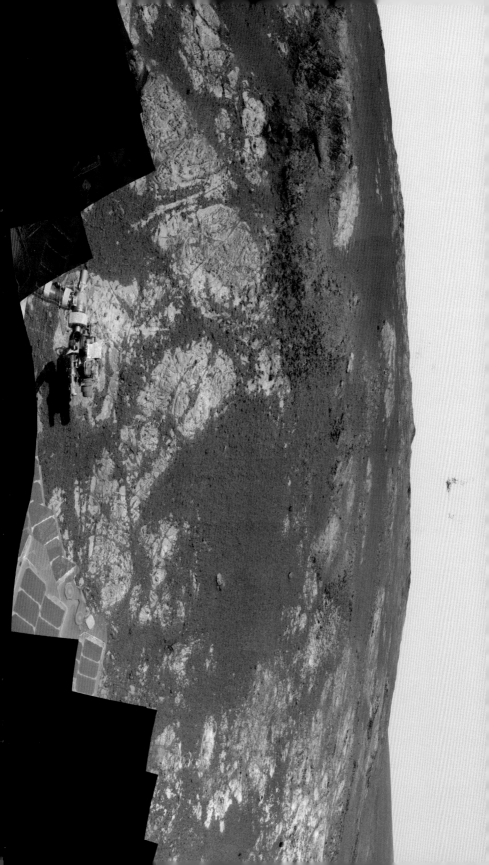

Bradbury
Landing

21 22
  24
3 16

26

29

27 39

40 41 42

45
43

North

Meters

0   25   50   100   150   200

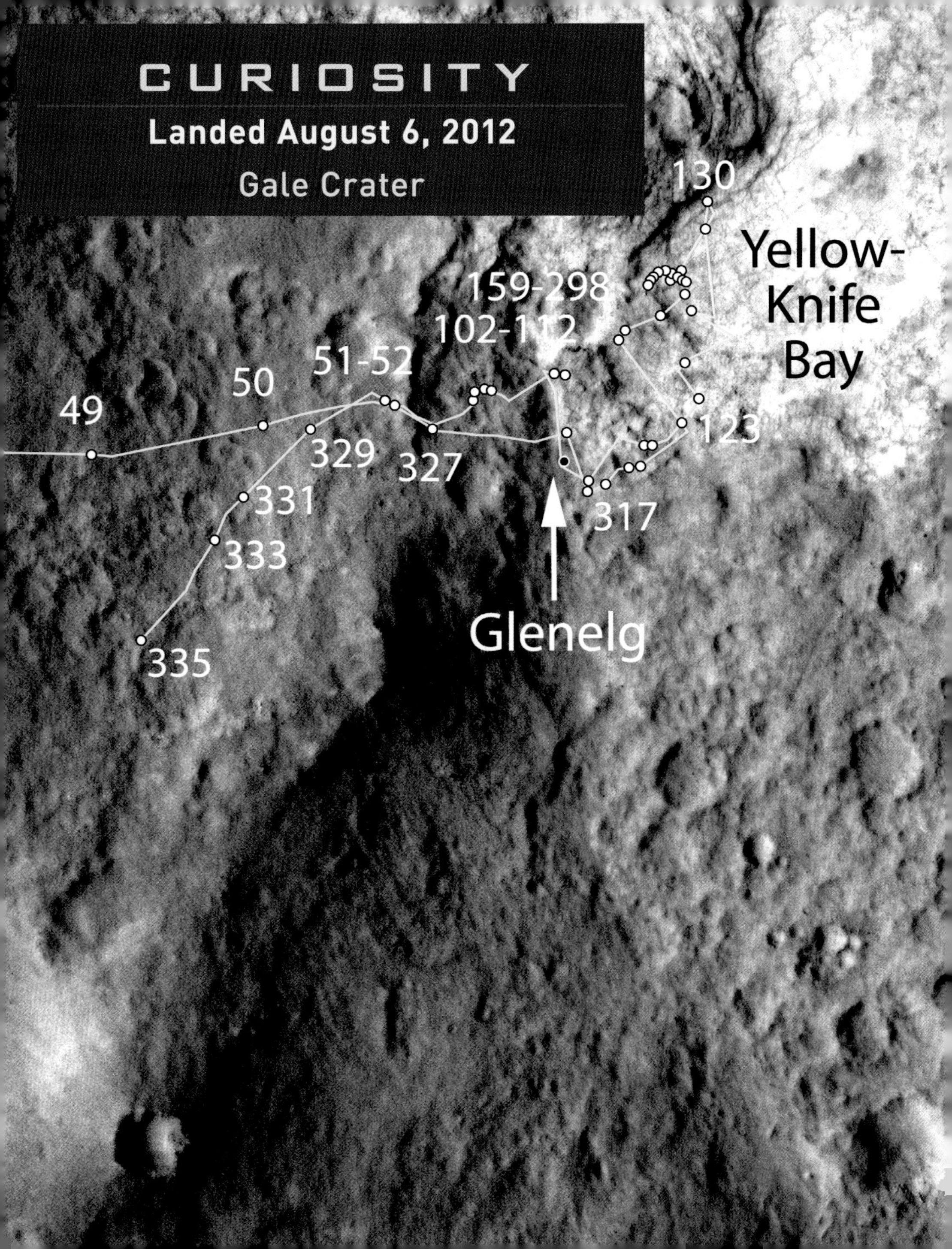

CURIOSITY
Landed August 6, 2012
Gale Crater

Yellow-Knife Bay

130

159-298
102-112

51-52

50

49

329

327

123

331

333

317

335

Glenelg

Mount Sharp

ABOVE *Curiosity's* Mastcam 100-mm-focal-length camera system obtains the highest resolution images yet acquired from any camera sent to the surface of Mars. This stunning image from sol 17 zooms in on some of the spectacular layering at the base of Mt. Sharp that compelled geologists to choose Gale crater as the rover's landing site. The pointy hill in the center is about 300 meters (1000 feet) across and 100 meters (300 feet) high.

OPPOSITE Although still nearly 10 kilometers (6 miles) from the rover at the time, the topography of Mt. Sharp—the 5-kilometer (3-mile)-high mound of layered sedimentary rocks in the middle of Gale crater—could still be resolved by *Curiosity's* cameras. This composite red–blue 3-D image was created by merging Mastcam images from sol 13 with Navcam images from a different location about 100 meters (300 feet) away on sol 32.

1 cm

Riverbed Conglomerate

ABOVE Similar evidence for water-rounded and transported rocks, pebbles, and gravel ("clasts") was found in the nearby region called "Link." Here, a sol 27 high-resolution Mastcam 100-millimeter camera image shows classic conglomerate structures within the interiors of these recently exposed rocks. Water transport is the only process capable of producing the rounded shape of clasts of this size. Scale bar is 1 centimeter (0.4 inches).

OPPOSITE This Mastcam sol 39 stereo image of the region called "Hottah" shows evidence of rounded rocks and pebbles cemented together in a kind of rock called a "conglomerate." On Earth, such features are evidence of past riverbeds—places where rocks were rounded and sorted as they were transported by water. *Curiosity's* first discovery was that similar vigorous rivers once flowed across the floor of Gale crater.

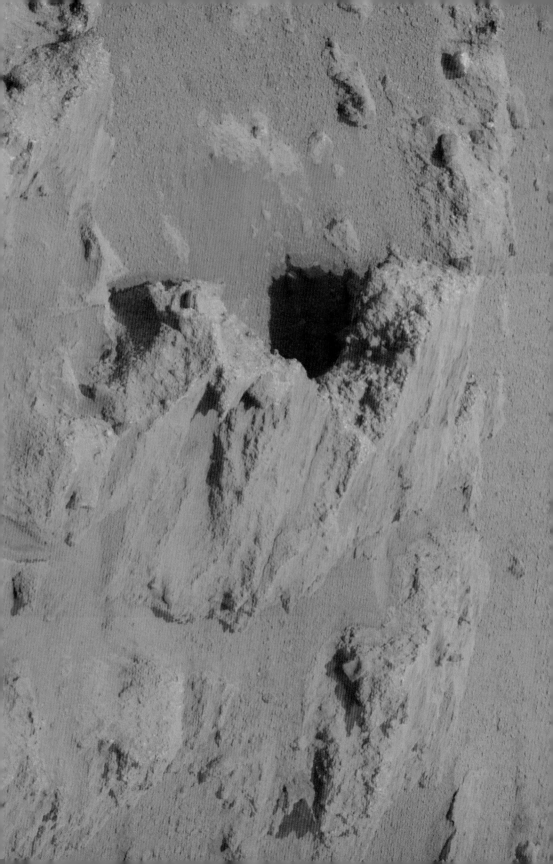

Self Portrait

ABOVE A portion of the sol 84 MAHLI color self-portrait centered on *Curiosity*'s remote sensing mast, the 1-meter (3-foot)-long boom housing the Mastcams (white squares), Navigation cameras (two on left, two on right), and the laser chemistry instrument called ChemCam (white box on top). Other instruments on the mast and deck include weather sensors sticking out from the mast, and the Mastcam calibration target/sundial at far right.

OPPOSITE *Curiosity*'s Mars Hand Lens Imager (MAHLI) camera can be positioned by the rover's arm so that it can look back and take self-portraits of the vehicle. This stereo version was taken from two different perspectives on sols 84 and 85, from a region called "Rocknest," the spot in Gale Crater where the mission's first scoop sampling took place. The scoop areas can be seen in the sand dune in front of the rover.

Armed for Science

ABOVE Another advantage of placing a high-resolution camera like MAHLI on the end of a nimble robotic arm is that self-inspection images of the rover can be taken from many different perspectives. This sol 34 "dogs-eye" view shows the rover's 6 wheels and belly, as well as the bank of four front Hazard avoidance wide-angle cameras (at top). Team members use images like this to assess driving obstacles and wheel wear-and-tear.

OPPOSITE The package of instruments on the turret at the end of the *Curiosity* rover's arm includes the MAHLI camera (centered here in this 3-D view from sols 84 and 85), a drill, a scooping and seiving tool, a brushing tool, and an X-ray spectrometer for measuring the chemistry of rocks and soils. The turret can spin 360 degrees to position the tools and is about the same size as the 1997 Mars Pathfinder *Sojourner* rover.

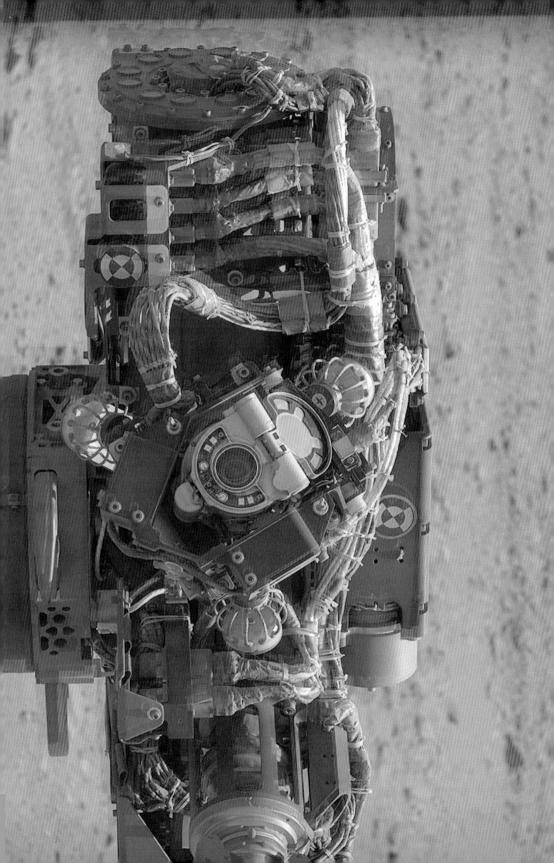

Gillespie Lake

ABOVE This *Curiosity* Mastcam sol 130 view shows dusty rocks and sandy ripples in the Gillespie Lake area of Yellowknife Bay. Mastcam and MAHLI images show the upper surfaces of these rocks to be coarse sandstones, perhaps the lithified result of the burial and later exhumation of ancient sand dune deposits. Underneath these could be water-deposited mudstones, although more analysis is needed to confirm.

OPPOSITE This stereo view of a sandy "beach" in front of the rock called "Gillespie Lake" was taken by the *Curiosity* rover's MAHLI instrument on sol 132, from a position near the lowest point within Yellowknife Bay. In addition to the fine sandy ripples, brighter veins can also be seen in this rock (near top center), indicating that ancient watery fluids may have precipitated minerals out of solution in this area.

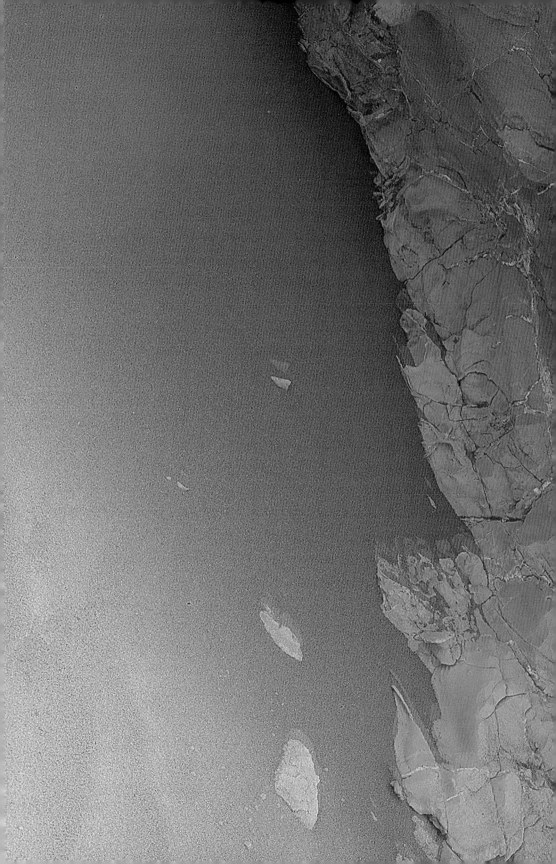

John Klein Veins

ABOVE Flatter mineral-filled veins as well as small, bright circular features informally called "mini bowls"—like these imaged by Mastcam on sol 133 in a 25-centimeter (10-inch)-wide area called Knorr—were also found in the region. Laser measurements of the chemistry and Mastcam infrared filter measurements suggested that these might be hydrated sulfates and that other water-bearing minerals might be found in the subsurface here.

OPPOSITE Choosing a site to perform the first robotic drilling on another planet presented a special challenge to the *Curiosity* rover team. Fortunately, the terrain within Yellowknife Bay was perfect: relatively flat, with interesting nearby geologic features, like these 0.5 to 1.0-centimeter (0.2 to 0.4-inch)-high raised ridges, suggestive of resistant mineral veins, some of which were accidentally crushed by the rover's wheels.

Yellowknife Bay

ABOVE Seen in more detail in this sol 170 Mastcam mosaic, the John Klein drill site (lower right) within Yellowknife Bay is an area that can be characterized as "polygonal terrain" because of its numerous polygon-shaped ground cracks. Reminiscent of some dry lake beds on Earth, many areas like this seen previously on Mars from orbital observations have shown evidence of ancient clay minerals formed in liquid water.

OPPOSITE The site ultimately chosen for the first drill activities by *Curiosity* was this flat, platy region called John Klein, named after the late *Curiosity* JPL Deputy Project Manager. Located within the bottom of the Yellowknife Bay depression, the region seen in this sol 166 Navcam stereo mosaic was speculated to have been where any liquid water in Gale crater stayed the longest, hopefully leaving telltale mineral evidence behind.

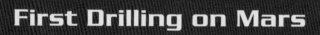

First Drilling on Mars

ABOVE This close-up sol 176 Mastcam inspection image shows the end of *Curiosity*'s robotic arm turret and the drill bit (center). The rounded cylinders sticking out past the drill provide a stable contact to the surface as the drill is slowly lowered. The flat, platy drill target rocks here are reddish and dusty, and it was unknown whether this generally oxidized nature of the surface rocks would extend into the subsurface.

OPPOSITE Finally ready to drill, *Curiosity* places its arm to drill down onto the John Klein rocks and make history. The flat and smooth nature of the drill target region—a requirement for the first full test of the drilling system on Mars—is obvious from this sol 174 Hazcam stereo view. After a test "mini-hole" was drilled on sol 180, a full drilling with capture of the drill tailings powder was made on sol 182.

1 cm

Gray Mars

ABOVE This sol 182 MAHLI image, taken from 25 centimeters (10 inches) above the surface, shows the mini-hole (left) and the first actual 6.4-centimeter (2.5-inch)-deep drill hole (right) into the John Klein target rocks. The gray drill powder indicates a less oxidized subsurface environment, and the clays detected in subsequent *Curiosity* analysis of the powder indicates that the water that once flowed here was fresh (neutral pH), rather than salty or acidic.

OPPOSITE Subsurface materials excavated from Yellowknife Bay turned out to be gray—poorly oxidized—rather than red. Another surprise was the discovery of up to 20 percent clay minerals in these drill samples. This MAHLI stereo view shows the 1.6-centimeter (0.6-inch)-wide second *Curiosity* drill hole in the Cumberland area. Not far from the first drilling site, part of the goal of drilling here was to confirm the results at John Klein.

Broken Stones

ABOVE Other unique rocks have been found along *Curiosity's* traverse so far. For example, this small 6-centimeter (2.4-inch)-wide rock imaged by Mastcam near the Bradbury Rise landing site on sol 27 shows a blue-black color overall, with lighter, whitish tones in patterns that may indicate the presence of distinct crystals of different material in the rock. Many more such "oddball" rocks are likely to be found along the rover's long drive to Mt. Sharp.

OPPOSITE *Curiosity* is a large vehicle—more than twice the size of the *Spirit* and *Opportunity* rovers. And so, as the rover drives along, it often churns the soil and crushes rocks. This Mastcam sol 188 stereo image shows the 12-centimeter (5-inch)-wide broken rock "Sutton Inlier" and the darker, sandy soils that were dug up as the rock was crushed. This is a rare example of a Mars rock with a grayish, subtly blue interior.

Web Links and other Resources

Websites

Spirit and Opportunity mission information: http://marsrovers.jpl.nasa.gov
Spirit 3-D images: http://marsrovers.jpl.nasa.gov/gallery/3d/spirit
Opportunity 3-D images: http://marsrovers.jpl.nasa.gov/gallery/3d/opportunity
Curiosity mission information: http://mars.jpl.nasa.gov/msl
Curiosity 3-D images: http://mars.jpl.nasa.gov/mars3d/
Pancam color imaging home page: http://pancam.sese.asu.edu/
The Planetary Society: http://www.planetary.org/
The Mars Society: http://www.marssociety.org/
Mars history: http://nssdc.gsfc.nasa.gov/planetary/chronology_mars.html
Mars facts: http://seds.lpl.arizona.edu/nineplanets/nineplanets/mars.html
Jim Bell's web site: http://jimbell.sese.asu.edu

Books

Bell, Jim. The Space Book. Sterling Publishing Company, Inc., 2006.

Bell, Jim, ed. **The Martian Surface: Composition, Mineralogy, and Physical Properties**. Cambridge University Press; 2008.

Bell, Jim. **Postcards from Mars: The First Photographer on the Red Planet**. Dutton, 2006.

Carr, Michael H. **The Surface of Mars**. Cambridge University Press, 2006.

de Goursac, Olivier. **Visions of Mars**. Harry N. Abrams, 2005.

Hartmann, William K. **A Traveler's Guide to Mars**. Workman Publishing Co., 2003.

Raeburn, Paul, and Matthew Golombek. **Mars: Uncovering the Secrets of the Red Planet**. National Geographic Press, 2000.

Squyres, Steven. **Roving Mars: Spirit, Opportunity, and the Exploration of the Red Planet**. Hyperion, 2005.

Magazine Articles

Bell, Jim. "Will Curiosity Find Life on Mars?", *Astronomy*, August 2012, 20-25.

Bell, Jim. "Have Brain, Must Travel," **Scientific American,** August 2007, 36-37.

Bell, Jim "Portraits from Mars," **Astronomy,** January 2007, 64-69.

Bell, Jim "The Red Planet's Watery Past," **Scientific American,** December 2006, 62-69.

Bell, Jim "Photographing Mars," **The Planetary Report,** November/December 2006, 12–18.

Bell, Jim "Backyard Astronomy from Mars," **Sky & Telescope,** August 2006, 40–44.

Bell, Jim "In Search of Martian Seas," **Sky & Telescope,** March 2005, 40–47.

Bell, Jim "Mineral Mysteries and Planetary Paradoxes," **Sky & Telescope,** December 2003, 34–40.

Bell, Jim "The Human Side of Mars Exploration," **The Planetary Report,** November/December 2003, 12-17.

Christensen, Philip R. "The Many Faces of Mars," **Scientific American,** July 2005, 32–39.

Petit, Charles W. "Making a Splash on Mars," **National Geographic,** July 2005.

INDEX

Note: **Bold** page numbers refer to captions and photographs.